W9-AEH-419

RELIABILITY ANALYSIS

and **PREDICTION** with

WARRANTY DATA

Issues, Strategies, and Methods

RELIABILITY ANALYSIS
and PREDICTION with
WARRANTY DATA

Issues, Strategies, and Methods

BHARATENDRA K. RAI
NANUA SINGH

CRC Press
Taylor & Francis Group
Boca Raton   London   New York

CRC Press is an imprint of the
Taylor & Francis Group, an **informa** business

CRC Press
Taylor & Francis Group
6000 Broken Sound Parkway NW, Suite 300
Boca Raton, FL 33487-2742

International Standard Book Number-13: 978-1-4398-0325-7 (Hardcover)

Visit the Taylor & Francis Web site at
http://www.taylorandfrancis.com

and the CRC Press Web site at
http://www.crcpress.com

Contents

SECTION I Need for Analysis and Prediction with Warranty Data, and Issues Involved

SECTION II   Strategies and Methods for Reliability Analysis with Warranty Data

SECTION III   Warranty Prediction

List of Figures

List of Tables

Notation

M_1	Warranty mileage limit
M_2	Warranty time limit
t	Months in service ($t = 1, 2, \ldots, M_2$)
N	Total number of vehicles in the field
V_t	Number of vehicles in the field up to t MIS
$N(t)$	Number of vehicles in the field without any claim at the beginning of MIS $= t$
a	Mileage value below which warranty data is artificially truncated
M_1-b	Mileage value above which warranty data is artificially truncated
n_t	Number of first claims in the mileage interval (a, M_1-b) miles at MIS $= t$ with $0 < a < b < M_1$
n_t^*	Number of first warranty claims at t
c_t	Number of left censored first claims at MIS $= t$
r_t	Number of vehicles that have not yet completed t MIS
W_t	Random variable denoting miles driven by a vehicle at MIS $= t$
$R(t)$	Reliability function at MIS $= t$
$h(t)$	Hazard function
$H(t)$	Cumulative hazard function

Acronyms

ANOVA	Analysis of variance
CCPR	Cumulative cost per repair
Cdf	Cumulative density function
CFR	Constant failure rate
CPU	Cost per unit
CRC	Customer reported concern
DFR	Decreasing failure rate
ICPR	Incremental cost per repair
IFR	Increasing failure rate
LN	Lognormal
MIS	Month(s) in service
MLE	Maximum likelihood estimator
MLP	Multilayer perceptron
OEM	Original equipment manufacturer
Pdf	Probability density function
R/1000	Repairs per thousand
RBF	Radial basis function
SND	Standard normal distribution

Preface

"What goes up, must come down" is an age-old saying. An exception to this seems to be warranty spending in the North American automotive industry. Warranty spending has reached close to double digits in billions of dollars per year, and all efforts to apply the brakes appear to be failing, time and again. In addition, there are market pressures from competitors to provide increased warranty coverage, leading to higher warranty costs and reduced profit margins.

Warranty costs affect companies financially, and are also a massive loss to society. Assuming a $50,000-a-year job, a loss of $8 billion implies an equivalent of 160,000 jobs lost. It also implies millions of dissatisfied customers. Why is this happening? Are engineers too busy looking in their rear mirrors to fix problems, with very little or no time left for looking forward?

It is time we ask where the problem lies. Are we doing our engineering right? Are we designing right, manufacturing right, and assembling right? Where did we lose a handle on such massive waste? The time has come to strike at the roots of the problem. Stephen Covey observes that from the same roots you will get the same fruits. If you want to change the fruits, you need to change the roots.

We believe that a major root cause, among others, is not utilizing a wealth of information hidden in the warranty data to make appropriate design-, manufacturing-, assembly-, or service-related improvements. It would not be an overstatement to say that every company strives to provide the best-quality products to their customers. For complex products such as an automobile, the majority of such efforts are directed at the R&D stage. At the development stage, activities such as concept/design failure mode and effects analysis, design verification planning and reporting, robust design experiments, etc., are performed to develop confidence that highly reliable and robust products have been developed and delivered. Various systems, subsystems, and components undergo prototype testing, life testing, and accelerated life testing for design verification and validation. In spite of employing the best quality and reliability practices, unexpected failures during the warranty period do occur and cost automobile companies billions of dollars annually in warranty alone.

Engineers seeking reliability/robustness improvements extensively use warranty data for feedback as they capture vehicle failures in true field conditions. In general, warranty data provide a rich source of information for modeling, analysis, and prediction to support strategic, tactical, and operational levels of decision making in automobile companies. However, the very nature of the warranty data makes such a task challenging on four counts: (1) availability of vehicle failure information is restricted to failures inside warranty limits (incompleteness), (2) failure reporting and diagnosis are not always accurate ("unclean"), (3) customer-rush leads to higher claims near warranty expiration limit (biased), and (4) increase in the warranty performance numbers when more data become available (warranty growth). Thus, to obtain valid and meaningful information/feedback from warranty data and to reduce errors at strategic, tactical, and operational levels of decision making, methods addressing such issues are useful.

This book aims to provide such methods and strategies for reliability analysis and prediction with warranty data that can help Six Sigma black belts and engineers to move in the right direction. It provides simple and practical approaches to address the issues identified. The book will be useful to engineers and Six Sigma black belts who extensively use warranty data to define and analyze field problems and seek guidelines for zeroing in on the root causes for warranty cost reduction. The book will also help quality and reliability engineers and professionals to be aware of the issues associated with warranty data and approaches to overcome them.

To achieve the stated goals, the book is divided into three parts. Part 1 provides background and introduction to reliability analysis and prediction using warranty data and highlights the issues involved. Part 2 gives the strategies and methods to obtain component-level nonparametric hazard rate estimates that provide important clues toward probable root causes and help reduce warranty costs. Part 3 of the book deals with prediction of the warranty performance. The methodologies covered help assess the impact of changes in warranty limits and forecast warranty performance.

Bharatendra K. Rai
Nanua Singh

Acknowledgments

We are thankful to John Koszewnik (Director, NA Diesel) and Katherine Franz (Diesel Quarterback Manager) of Ford Motor Company for taking interest in our work and supporting the research position in the company. We are grateful to Bob Knecht (Manager, Warranty Analysis and Administration) for allowing the use of the previous model-year warranty data for research purposes by masking information such as model year, vehicle line, repair type, subsystem, part name/number, etc. We would also like to acknowledge the thought-provoking ideas of Tim Davis and Vasiliy Krivtsov of Ford Motor Company during their reliability and robustness seminars, which motivated the authors' interest in this field of research.

We thank Kurt Schieding of Bank of America (formerly with Ford Motor Company) for continuous encouragement and support during the research work. We also like to thank section supervisors Lisa Fullerton, Dianna Ball, Gary Mazzella, and Ramana Divakaruni of Ford Motor Company for providing opportunities to work on research-related problems and for taking interest in the research work.

We are also very thankful to Cindy Renee Carelli, Jim McGovern, Jennifer Ahringer, and Ashley Gasque of the Taylor & Francis Group for their continued support throughout this book project.

BKR
NS

The Authors

Bharatendra K. Rai, Ph.D., is an assistant professor of business statistics at the Earle P. Charlton College of Business, Wayne State University, Detroit. He received his Ph.D. in industrial engineering from Wayne State. His two master's degrees include specializations in quality, reliability, and operations research from the Indian Statistical Institute and statistics from Meerut University, India. He has vast experience in the area of applied statistics. This includes over seven years of consulting and training experience with various Indian industries such as automotive, cutting tool, electronics, food, pharmaceutical, software, chemical, defense, etc., in the areas of statistical process control, design of experiments, quality engineering, problem solving tools, Six Sigma, and quality management systems. He also has over five years worth of extensive research and work experience at Ford Motor Company in the area of quality, reliability, and Six Sigma.

Dr. Rai's research interests include applications in multivariate diagnosis/pattern recognition and data mining, developing meta models using computer experiments, prediction of unexpended warranty costs, and field performance studies from large warranty datasets. His research publications have appeared in journals such as *IEEE Transactions on Reliability, Reliability Engineering and System Safety, Quality Engineering, International Journal of System Sciences, International Journal of Product Development,* and *Journal of System Science and System Engineering.* He has presented his research work at professional meetings and conferences such as the SAE World Conference, INFORMS annual meetings, the Industrial Engineering Research Conference, American Society for Quality's (ASQ) Annual Quality Congress, Taguchi's Robust Engineering Symposium, and the Canadian Reliability and Maintainability Symposium. He has also coauthored book chapters in the area of applied statistics.

Dr. Rai has won awards for excellence and exemplary teamwork at Ford Motor Company for his contributions in the area of applied statistics. He also received an Employee Recognition Award by Ford Asian Indian Association (FAIA) for his Ph.D. dissertation in support of Ford Motor Company. He is certified as an ISO 9000 lead assessor from the British Standards Institute and as an ISO 14000 lead assessor from Marsden Environmental International, and has been designated a Six Sigma black belt by the American Society for Quality. He can be contacted at:

Email: brai@umassd.edu
Phone: (508) 999-6434
Office: CCB 326

Nanua Singh, Ph.D., is the president of RGBSI (Rapid Global Business Solutions, Inc.). Dr. Singh sets the strategic direction and goals for the global growth of RGBSI. His main objective is to provide innovative solutions to global clients by leveraging the talent and years of expertise that RGBSI resources possess. RGBSI is an American company, founded in March 1997 with its headquarters in Madison

Heights, Michigan, and offices in Manhattan (New York), Bangalore (India), Shanghai (China), Dusseldorf (Germany), and Toronto (Canada).

An academician-turned-entrepreneur, Dr. Singh has published three books in engineering, with another forthcoming. He has supervised over twenty Ph.D.s and over sixty master's theses, and published and presented over 130 research papers in a variety of international journals/conferences.

His career highlights include being a professor at the prestigious India Institute of Technology (IIT) Delhi (India), head of the Department of Industrial Engineering at the University of Windsor (Canada), and tenured professor of manufacturing engineering at Wayne State University in Detroit. Dr. Singh spent one year on his sabbatical in the Department of Industrial and Operations Engineering at the University of Michigan, Ann Arbor, Michigan.

Section I

Need for Analysis and
Prediction with Warranty
Data, and Issues Involved

1 Reliability Studies with Warranty Data
Need and Issues

No problem can stand the assault of sustained thinking.

—Voltaire

OBJECTIVES

This chapter explains the following:

- The role of field data in continuous improvement of products
- Key factors that influence warranty cost
- Different levels of decision making with warranty data
- Importance of hazard function obtained from warranty data
- Definitions and concepts related to warranty
- Issues with analysis involving warranty data and their implications

OVERVIEW

This chapter essentially provides the necessary background and motivation for the strategies and methods presented in the book. Section 1.1 describes the role of field data in continuous improvement of products and major factors that influence the warranty cost. Section 1.2 provides three levels of decision making with warranty data and their likely impact in terms of time and money, with examples. Section 1.3 emphasizes the importance of the hazard function obtained from warranty data in providing directions for reliability and robustness improvement. Section 1.4 explains the issues associated with warranty data that make modeling and analysis from warranty data challenging. Section 1.5 discusses existing methods and their limitations, and Section 1.6 outlines the scope and objective of this book. Section 1.7 covers some basic warranty concepts, and Section 1.8 outlines the remaining chapters of the book.

1.1 CONTINUOUS IMPROVEMENT AND FIELD DATA

Most companies strive to provide the best-quality products to their customers. Most such efforts take place at the research and development stage. Taguchi (1992) and Wu and Wu (2000) point out that once drawings and specifications are completed

3

and production processes selected, product quality is almost fully determined. Then there is little the production engineers can do to further improve product quality. At the development stage, various activities, including concept/design failure mode and effects analysis, design verification planning and reporting, and robust design experiments, are performed. The effectiveness of these activities in developing reliable and robust products is very often judged through laboratory life testing.

An automobile with thousands of parts and interactions among them is a highly complex product. It makes detailed testing/analysis during product development, manufacturing, and assembly a prohibitively enormous, if not infeasible, task. Thus, when the vehicle is put on the market, the likelihood of unexpected poor quality and reliability resulting in high warranty costs is not uncommon. Success during product development, manufacturing, and assembly is often judged by lack of quality and reliability problems when the vehicle is in the field.

Traditionally, reliability has been defined as follows (Andrews and Moss 2002):

> The probability that an item (component, equipment, or system) will operate without failure for a stated period of time under specified conditions.

The word *specified* in the preceding definition relates better to laboratory testing, where the test conditions can be specified. However, the specified conditions for the item may not match the actual conditions encountered during its field usage. For example, the specified conditions of laboratory testing of a steering wheel assembly may not include the fact that the users may use it as a handle to get inside the vehicle. Meeker and Escobar (2004) suggest use of the word *encountered* as more appropriate in place of the word *specified* in the definition of reliability. However, during laboratory testing, it is almost infeasible to replicate potentially every encountered usage condition. Therefore, success from laboratory life testing alone does not give design engineers full confidence and feedback about the field performance.

It has been noted that field data provide more reliable information about the life distribution compared to laboratory data (Suzuki 1995; Oh and Bai 2001). Field data capture actual usage profiles and the combined environmental exposures that are difficult to simulate in the laboratory. Occasionally, the automobile companies use fleets of production vehicles to obtain rapid feedback on the nature of field failures (Inman and Gonsalvez 1998). Due to readily accessible data from the limited number of vehicles in the fleet, the modeling and analysis of such data does not pose a serious problem. However, in automotive companies, access to field data is mainly through warranty data. Hence, designers and engineers working on current and forward model vehicles also look forward to warranty data for more accurate feedback on design. Engineers and Six Sigma black belts use warranty data to identify opportunities for warranty cost reduction through appropriate design, manufacturing, or service fix. Some of the major factors that influence overall warranty cost of a new vehicle are depicted in Figure 1.1 using a parameter diagram (Phadke 1989).

Murthy and Djamaludin (2002) note that the decisions and actions during design, manufacturing, and assembly determine the inherent reliability of the product. A product that performs well even in the presence of noise factors is said to be robust. It is expected that the better the control over design, manufacturing, and assembly variables,

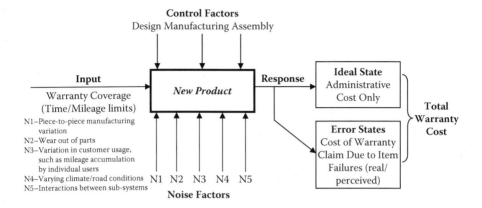

FIGURE 1.1   Factors influencing warranty cost.

the lower will be the warranty cost. In an ideal situation, when the control over design, manufacturing, and assembly variables is perfect and the effect of noise factors is negligible, the total warranty cost will only be the administrative cost of maintaining the warranty system (Figure 1.2a). However, in reality, noise factors do play a significant role and the error states lead to a major portion of the total warranty cost (Figure 1.2b). For a given control and noise factor, the change in warranty coverage for new vehicles provided by the manufacturer significantly influences total warranty cost.

1.2   THREE LEVELS OF DECISION MAKING WITH WARRANTY DATA

In the past, when major North American automobile companies made good profits, detailed warranty modeling was probably not that necessary. But today, when on the one hand major automakers spend billions of dollars annually in warranty cost and on the other they have substantially reduced profit margins, warranty modeling is no more a luxury but has become a necessity. Petkova et al. (1999) note that if such information is well analyzed and communicated, the recurrence of old problems in new products will be drastically reduced and so will the expenses on recalls, repairs, warranties, and liabilities.

Modeling, analysis, and prediction using warranty data are generally undertaken to support three levels of decision making in a company, that is:

- Strategic level
- Tactical level
- Operational level

Strategic-level decisions have a long-term impact on the company, both financially and in terms of time (see Figure 1.3). For example, increasing the warranty coverage for a vehicle from 3 year/36K mi to 5 year/50K mi may impact the company for 10 or more years. Although strategic-level decisions are less frequent than tactical or

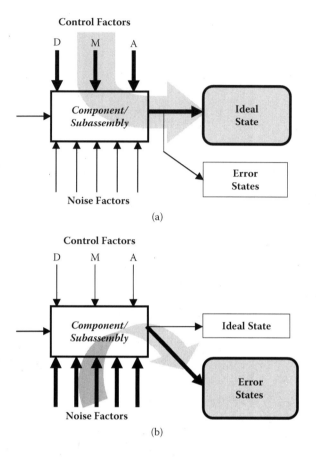

FIGURE 1.2   The influence of control factors and noise factors on total warranty cost: (a) control factors influence ideal response, (b) noise factors lead to occurrence of error states.

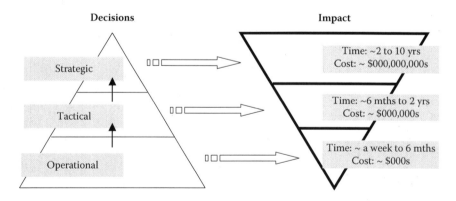

FIGURE 1.3   Strategic, tactical, and operational decisions using warranty data and their impact.

operational decisions, the cost impact of such decisions could easily run into billions of dollars for major automakers. Also, once a decision to increase the warranty coverage is implemented, it may become difficult to roll back as it may give a negative signal to customers. Due to the magnitude of impact involved, the decision makers at this level are usually the senior management. One of the major noise factors at the strategic level of decision making is action by a competitor.

Tactical-level decisions may impact a company for durations varying from 6 months to 2 or 3 years. Planning the supply of replacement parts based on the component failure predictions over the next 1 or 2 years is an example. It is important that tactical-level decisions align with strategic-level decisions. The decision makers at the tactical level are generally the middle management of a company.

Operational-level decision making with warranty data has the shortest time duration, and may vary from a week to about 6 months. An example is estimating the expected cost savings from an improvement project using warranty data. Operational-level decisions are generally aligned to the tactical- and strategic-level decisions, and are made by engineers and Six Sigma black belts.

Some of the important benefits of modeling, analysis, and prediction from automobile warranty data sets, their application area/activity, and linkage to level of decision making are summarized in Table 1.1.

1.3 THE ROLE OF HAZARD FUNCTION IN RELIABILITY AND ROBUSTNESS IMPROVEMENTS

Murthy and Djamaludin (2002) indicate that the purpose of warranty is to establish liability between the two parties (manufacturer and buyer) in the event that an item fails. An item is said to fail when it is unable to perform satisfactorily its intended function when properly used. Although existence of warranty data is primarily owing to financial reasons, it is extensively used in reliability studies to provide appropriate feedback for continual design improvements.

In field reliability studies, the hazard function denoted by $h(t)$ plays an important role. For a continuous nonnegative random variable T representing lifetimes of individual items, the hazard function (Nelson 1982; Lawless 1982; Crowder et al. 1991; Meeker and Escobar 1998) is defined as follows:

$$h(t) = \lim_{\Delta t \to 0} \frac{P(t \leq T \leq t + \Delta t \mid T \geq t)}{\Delta t} \tag{1.1}$$

where Δt denotes a very small time interval. The hazard function specifies the instantaneous failure rate at time t given that an item has survived until t. Blischke and Murthy (2000) note that hazard rate characterizes the effect of age on an item failure more explicitly than does the failure distribution or the density function. Hazard plots provide estimates of the distribution parameters, the proportion of units failing by a given age, percentiles of the distribution, the behavior of the failure rate as a function of age, and conditional failure probabilities for units of any age (Nelson 2000). It also enables determination of the bathtub curve as shown in Figure 1.4.

The shape of the hazard plot provides important directions to the engineers working on improvement projects. Decreasing failure rate pattern indicates influence of mainly manufacturing/assembly variables, a constant failure rate pattern indicates effect of random causes, and increasing failure rate pattern generally indicates wearing-out failures. Davis (2003) further notes that the noise factors have different impacts on the failure rate of the components. Decreasing failure rate pattern is

TABLE 1.1

Linkage among Levels of Decision Making, Benefits of Modeling, Analysis, and Prediction from Warranty Data and Application Area/Activity

Level of Decision Making	Decision Makers	Benefits of Modeling, Analysis, and Prediction from Warranty Data	Application area/activity		
			Design/ Mfg./ Assembly	Continuous Improvement/ Six Sigma	Warranty Management
Strategic	Senior managers	1. Design of warranty programs			✓
		2. Data bank about failure modes and their relation to the environmental and usage conditions	✓	✓	
		3. Relationship among test data at development stage, inspection results of production, and field performance	✓	✓	
Tactical	Middle managers	4. Early warning/detection of wrong design, production process, parts or materials	✓	✓	
		5. Planning the supply of replacement parts			✓
		6. Assessment and refinements of reliability predictions		✓	✓
Operational	Engineers and Six Sigma black belts	7. Comparison of performance before and after design, manufacturing, or service fix	✓	✓	
		8. Selection and justification of engineering design improvement projects		✓	
		9. Prediction of unexpended warranty		✓	✓

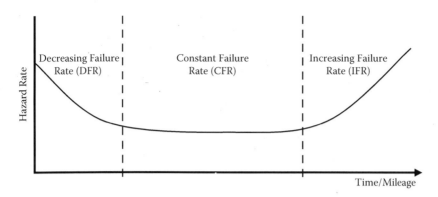

FIGURE 1.4   A typical bathtub curve.

mainly influenced by the N1 category of noise factors (Figure 1.5a), whereas increasing failure rate pattern is largely a result of the N2 category (Figure 1.5b). Noise factors in categories N3, N4, and N5 jointly or individually lead to observations of constant failure rate (Figure 1.5c). Thus, the failure rate pattern obtained from a hazard plot can provide important clues pertaining to, and insights into, making a component/subsystem robust against the key noise factors. Table 1.2 provides a summary of the relationship between noise factors and failure pattern on the hazard plot.

1.4   WARRANTY DATA: NOT ALWAYS PERFECT FOR STATISTICAL ANALYSIS

Hazard plots obtained using warranty data may not capture the entire life cycle of a component or a subsystem because of unavailability of failure information beyond the warranty period. Accordingly, a hazard plot may result in one of the seven failure rate patterns depicted in Figures 1.6a–g. The ability to arrive at one of these seven hazard plots is key to initiating and implementing appropriate corrective and preventive measures through a focused root cause analysis, leading to reduced field failures and in turn to reduced warranty cost.

Notwithstanding the usefulness of warranty data, the following are certain major shortcomings:

- **Incompleteness:** In North America, automobile warranties are generally two-dimensional and are stated in terms of time (in months or years) and mileage. They expire when any of the two limits is crossed. For failures reported during the warranty period, numerous vehicle- or failure-related details such as model year, date of purchase, date of repair, mileage at failure, failure mode, customer/technician comments, etc., are recorded in the warranty database. However, no record is available for the vehicles that have either not yet failed during the warranty period or failed outside the warranty period. This makes the automobile warranty data sets incomplete in assessing the failure rates.
- **Uncleanliness:** Inaccurate reporting of failures, incorrect causal part attribution, etc., often renders warranty data unclean for field reliability studies.

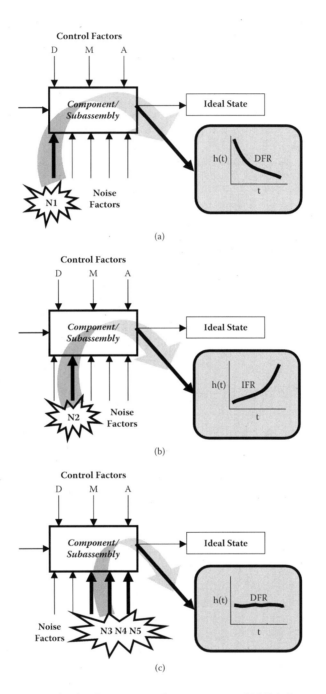

FIGURE 1.5 Impact of noise factors on total warranty cost: (a) N1 influences decreasing failure rate pattern on the hazard plot, (b) N2 influences increasing failure rate pattern on the hazard plot, (c) N3, N4, and N5 influence constant failure rate pattern on the hazard plot.

TABLE 1.2

Relationship between Noise Factors and Failure Pattern

Noise Factor Category	Description	Expected Failure Pattern on a Hazard Plot
N1	Piece-to-piece manufacturing variation	Decreasing failure rate (DFR)
N2	Wearing out of parts with time and usage	Increasing failure rate (IFR)
N3	Variation in customer usage	Constant failure rate (CFR)
N4	Varying climate/road conditions	Constant failure rate (CFR)
N5	Interactions between subsystems	Constant failure rate (CFR)

As unclean warranty data could potentially hide inherent failure patterns, warranty claims are often screened before performing field reliability studies.

- **Bias:** Sometimes customers experiencing soft failures in their vehicle (failures that result in degraded performance, but the vehicle can still be operated) delay reporting of the failure until the coverage is about to expire. This leads to relatively higher number of claims closer to the warranty expiration limit and introduces a bias in the warranty data set. Note that some soft failures may get reported with other failures or may never get reported during the warranty period.

- **Warranty growth or maturing data phenomena:** Two commonly used warranty performance measures in automobile companies are repairs per 1000 vehicles in the field (R/1000) and cost per unit or vehicle (CPU). The value of these warranty performance indicators at a specific month in service (MIS) changes with time. As this change is often in the increasing direction, this phenomenon is termed *warranty growth*. It makes the forecasting of warranty performance challenging.

When such warranty data are used for obtaining feedback on component reliability, it may potentially mislead product or reliability engineers with a distorted picture of the reality. Such information can be misleading, especially at a time when companies are spending a great deal of time and money to reduce warranty cost, and can lead to the following errors (Figure 1.7):

- **Error at strategic level:** Choosing a warranty policy that proves to be costlier than expected.

- **Error at tactical level:** The improvement projects with the potential to have significant influence on warranty cost are either not identified or are identified very late, when significant warranty costs have already been incurred.

- **Error at operational level:** An inaccurately assessed increasing failure rate pattern instead of the actual decreasing or constant failure rate may lead engineers to work on costly and unwarranted design changes.

The implication of such errors could easily run into millions of dollars for a company. The strategies and methods in this book help one obtain valid and meaningful

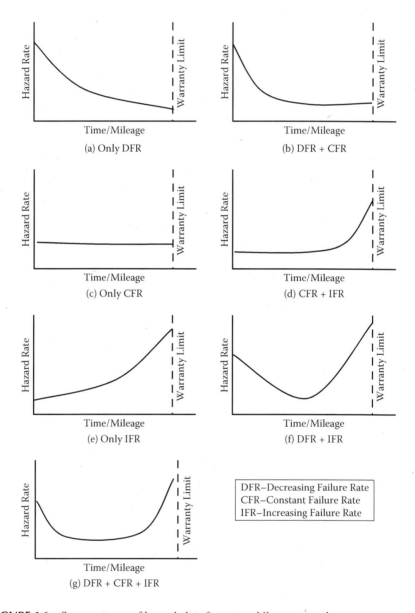

FIGURE 1.6   Seven patterns of hazard plots for automobile warranty data.

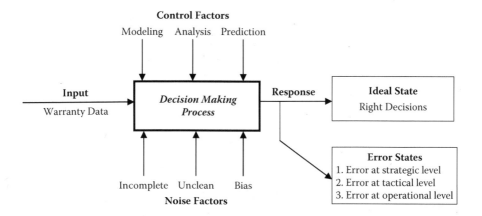

FIGURE 1.7   Parameter diagram for the decision-making process using warranty data.

information and feedback from the warranty data, which enables sound decision making at strategic, tactical, and operational levels.

1.5 EXISTING RESEARCH WORK AND CERTAIN LIMITATIONS

The mid-1980s to early 1990s saw growing interest by several researchers to address certain general issues associated with the modeling and analysis of warranty data (Suzuki 1985a, 1985b; Kalbfleisch and Lawless 1988; Kalbfleisch et al. 1991; Hill et al. 1991; Singpurwalla and Wilson 1993; Chen et al. 1996). However, over the years there has been an increase in efforts to address more specific and practical issues dealing with studies involving warranty data (Lawless et al. 1995; Lu 1998; Krivstov et al. 2002; Majeske 2003; Stephens and Crowder 2004; Rai and Singh 2004a; Elkins and Wortman 2004). Some of the issues related to warranty data that were major issues earlier, such as reporting delays, are now virtually eliminated with modern reporting systems (Wu and Meeker 2002). Also, with rapid advances in information technology, maintaining large warranty databases and utilizing it for various kinds of studies with proper modeling and analysis has become more practical.

The increase in warranty coverage from 12/12 in the 1980s to 7/70 now by certain manufacturers has made more and more failure data available through warranty databases. However, with the increasing size of warranty databases and human factors involved in data recording, there are growing concerns about the data quality for quality and reliability studies. Several factors related to customers and the technicians who carry out the diagnosis influence reported failures. Although issues of complex censoring and truncation associated with warranty data has been addressed by several researchers, inaccurate reporting of failures and other related issues highlighted in this book, such as customer-rush near warranty expiration limit, are yet to be addressed.

Another limitation of existing methods involves warranty cost studies. The total warranty cost for a given model-year vehicle can be arrived by estimating the number of failures reported within the warranty period and the cost per repair. Chen and Popova (2002) and Baik et al. (2004) consider warranty cost modeling for two-dimensional warranties with fixed cost per repair. However, cost per repair may vary over the life cycle of a vehicle.

1.6   THE SCOPE AND OBJECTIVE OF THE BOOK

The interplay among manufacturer quality, warranty system, and statistical methods is shown in Figure 1.8. The elements involved in the interplay among the three areas are as follows:

a. $A \cap \bar{B} \cap \bar{C}$—Best practices in design, manufacturing, and assembly such as axiomatic design, finite element analysis, poka-yoke or mistake-proofing, single minute exchange of dies, etc.

b. $\bar{A} \cap B \cap \bar{C}$—Knowledge base of statistical methods.

c. $\bar{A} \cap \bar{B} \cap C$—Warranty policies, government regulations, market conditions; processes for operating and maintaining warranty database.

d. $A \cap B \cap \bar{C}$—Application of statistical methods in laboratory testing, design verification and validation, accelerated testing, process control, etc.

e. $A \cap \bar{B} \cap C$—Design for serviceability and technician diagnostics.

f. $\bar{A} \cap B \cap C$—Application of statistical methods to monitor supplier performance using warranty data.

g. $A \cap B \cap C$—Feedback on manufacturer quality from warranty data using statistical/reliability analysis, warranty prediction, warranty cost analysis, etc.

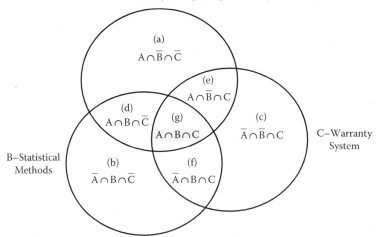

FIGURE 1.8 Interplay among manufacturer quality, warranty system, and statistical methods.

The interface of statistical methods, manufacturer quality, and warranty system, represented by $A \cap B \cap C$, is the focus area of this book. The book further focuses on component-level modeling, analysis, and prediction from the warranty data sets.

One of the main objectives of a component reliability study from warranty data is to support an engineer's efforts in getting an insight into the phenomena under study. In the words of Dr. C. R. Rao (recent recipient of National Medal for Science by the President of United States of America):

> Statistical methodology is a process by which we analyze data to provide insight into the phenomenon under investigation rather than a prescription for final decision.

Some of the questions that this book helps to answer include the following:

1. What are the key characteristics of automobile warranty data and how do they impact field reliability studies?
2. What role does the hazard analysis play in providing engineers with a direction for root cause analysis?
3. How can hazard rate estimates be arrived at in the presence of incomplete and unclean data available from warranty data sets?
4. How does the methodology differ in the presence or absence of knowledge/ data on mileage accumulation rates in the vehicle population?
5. How are the hazard rate estimates obtained when the warranty data set is influenced by customer-rush near the warranty expiration limit?
6. How can the impact of changing time and/or mileage limits in warranty coverage of a vehicle be assessed?

1.7   WARRANTY CONCEPTS

Blischke and Murthy (1994) note that until the 1950s, warranty studies were mainly carried out by researchers in the legal profession and, thus, such articles appeared only in the law journals. Subsequently, warranty studies attracted researchers from diverse fields such as economics, accounting, management, marketing, engineering, operations research, and statistics. Singpurwalla and Wilson (1993) define warranty as a contractual agreement that requires the manufacturer to rectify a failed item, either through repair or replacement, should failure not be attributed to reckless use, and should it occur during a period specified by the warranty. For a vehicle under warranty coverage, customer-initiated vehicle repairs are carried out by an authorized service or repair center. This section briefly covers some of the key warranty concepts and terminology.

1.7.1   ONE- AND TWO-DIMENSIONAL WARRANTY COVERAGE

Automobile warranty coverage is usually of two types: one dimensional and two dimensional. One-dimensional warranty coverage is almost always based on time or age. In contrast, two-dimensional warranty coverage includes both time and usage. Blischke and Murthy (1994) provide an example of one-dimensional warranty for a

1930-model-year vehicle where the manufacturer provided 90-day warranty coverage for new vehicles. However, for any kind of repair or replacement, a consumer needed to deal directly with the component manufacturers at that time. Today, many vehicles in the European market carry warranty coverage of one year with unlimited mileage. In North America, two-dimensional warranty coverage involving 3 years or 36K mi (or simply 3/36K), whichever comes first, are more prevalent. Some manufacturers also provide 5/100K or 7/70K warranty coverage on their vehicles.

1.7.2   BASE AND EXTENDED WARRANTY

Base warranty coverage is the original warranty coverage provided by a manufacturer. Such coverage is offered at no additional cost as the cost of base coverage is built into the selling price of a vehicle. Auto manufacturers also provide an option of purchasing an extended warranty coverage that comes into effect after the expiration of the base coverage. Extended coverage is an optional warranty and is not tied to the sale process. However, the buyers need to pay an extra premium to purchase such an extended coverage.

Irrespective of the basic and extended warranty coverage, certain states mandate automakers to provide a specified coverage on certain subsystems or components. For example, cars sold in California require certain components of the emission control systems in a vehicle to be covered for 7/70K.

1.7.3   FAILURE MODES, CAUSES, AND SEVERITY

Blischke and Murthy (2000) define failure as the termination of the ability of an item to perform a required function. Associated with every failure are the failure modes, their causes, and the severity that we describe briefly in the following text:

Failure modes: Hoyland and Rausand (1994) define failure mode as the effect by which a failure is observed on the failed item. Blischke and Murthy (2000) classify failure modes into two: intermittent failures and extended failures. Intermittent failures are those that may be difficult to duplicate when reported at a service center. Extended failures are those that can be observed until corrected. Extended failures are further classified into two: complete (or hard) failure and partial (or soft) failures. As the name suggests, a complete failure leads to a total loss of function, whereas a partial failure leads to partial loss of function. The occurrence of both complete and partial failures may be either sudden or gradual.

Failure causes: Failure cause pertains to the circumstances during design, manufacture, or use that have led to a failure (Blischke and Murthy 2000). Failure causes may include the following:
   • Design failure
   • Weakness failure
   • Manufacturing failure
   • Aging failure

- Misuse failure
- Mishandling failure

From among these failure causes, misuse and mishandling causes of failures may not be covered by manufacturer warranty.

Failure severity: Blischke and Murthy (2000) define failure severity as the impact of the failure mode on the system as a whole and on the outside environment. MIL-STD 882 (1984) classifies failure severity into four: catastrophic, critical, marginal, and negligible. A catastrophic failure may result in death or total system loss. Critical failures may result in severe injury or major system damage. A marginal failure is one that results in minor injury or minor damage to the system, whereas a negligible failure may result in less than minor injury or system damage.

1.7.4 ROLE OF WARRANTY

Blischke and Murthy (2000) point out that warranties serve different purposes for the buyer and manufacturer. From a buyer's point of view, the role of warranty is mainly to provide protection in the case of vehicle failure. It assures the buyer that a vehicle which does not perform as expected or fails even though used properly will be either repaired or replaced at no or minimal cost to the buyer. Another way in which the warranty influences a buyer is informational, as sometimes buyers may infer that a vehicle with higher warranty coverage is more reliable and robust.

From the manufacturer's point of view, too, one of the main roles of the warranty is to provide protection. The terms and conditions of the warranty protect manufacturers against any misuse. Another important role of the warranty from the manufacturer's point of view is promotional. As buyers often view products with longer warranty coverage to be more robust and reliable, it is frequently used as an advertising tool. Sometimes, competition may force companies to provide more warranty coverage to gain buyer confidence.

1.7.5 IMPORTANT FUNCTIONS IN RELIABILITY STUDIES FROM WARRANTY DATA

Five important functions in studies involving warranty data are distribution function denoted by $F(t)$, probability density function denoted by $f(t)$, reliability function denoted by $R(t)$, hazard function denoted by $h(t)$, and cumulative hazard function denoted by $H(t)$. Table 1.3 gives the relationships among these five functions.

Although some of the formulae in Table 1.3 imply continuous failure data, extension to discrete data is straightforward.

1.7.6 SIX SIGMA AND WARRANTY DATA

The Six Sigma methodology is a popular approach in industries to make significant improvements in products and processes. A typical Six Sigma improvement project is led by a black belt and has five well-defined phases: define, measure, analyze, improve, and control (Harry and Schroeder 2000). In the *define* phase, the processes that contribute to the functional problem are defined. Functional problems could be

TABLE 1.3

Relationship between Functions $F(t)$, $f(t)$, $R(t)$, $h(t)$, and $H(t)$

Function	$F(t)$	$f(t)$	$R(t)$	$h(t)$	$H(t)$
$F(t)$	—	$\displaystyle\int_0^t f(u)du$	$1 - R(t)$	$1 - e^{-\int_0^t h(u)du}$	$1 - e^{-H(t)}$
$f(t)$	$\dfrac{d}{dt}F(t)$	—	$-\dfrac{d}{dt}R(t)$	$h(t)e^{-\int_0^t h(u)du}$	$e^{-H(t)}\dfrac{d}{dt}H(t)$
$R(t)$	$1 - F(t)$	$\displaystyle\int_t f(u)du$	—	$e^{-\int_0^t h(u)du}$	$e^{-H(t)}$
$h(t)$	$\dfrac{dF(t)/dt}{1-F(t)}$	$\dfrac{f(t)}{\displaystyle\int_t f(u)du}$	$-\dfrac{d}{dt}\ln R(t)$	—	$\dfrac{d}{dt}H(t)$
$H(t)$	$-\ln(1-F(t))$	$-\ln\left(\displaystyle\int_t f(u)du\right)$	$-\ln(R(t))$	$\displaystyle\int_0^t h(u)du$	—

of three types: product problems, service-related problems, and transactional problems. In the *measure* phase, the process performance and capability of each process is measured and quantified. This further leads to the identification of critical-to-quality (CTQ) characteristics. In the *analyze* phase, the data is analyzed to assess prevalent patterns and trends that in turn help identify potential root causes. In the *improve* phase, the key product/service characteristics are improved using methods such as design of experiments. And in the *control* phase, the key variables are controlled within their operating limits over time for the improved process.

The Six Sigma black belt projects, which aim to reduce warranty costs for the current and forward model-year vehicles, very often require modeling and analysis using warranty data. Define and measure phases are the two main phases that benefit from such an analysis, which helps quantify projected levels for reliability and cumulative hazard. After the improve phase, once the design/service changes are implemented, the latest warranty data help judge the effectiveness of the improvement actions.

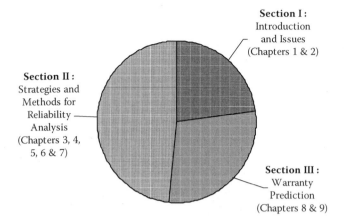

FIGURE 1.9 Organization of the book.

1.8 ORGANIZATION OF THE BOOK

This book is organized into nine chapters grouped in three sections, as shown in Figure 1.9:

Section I of the book has two chapters, including Chapter 1, which is an introduction to studies involving warranty data and highlights the issues involved. Chapter 2 describes the characteristics of automobile warranty data and provides insight into the major issues involved in modeling and analysis of warranty data. It also outlines broad strategies for obtaining hazard rate estimates from the warranty data.

Section II constitutes the largest portion of the book, and deals with strategies and methods for reliability analysis with warranty data. Chapter 3 outlines different methods for analyzing warranty data along with a broad strategy for arriving at hazard rate estimates. Chapter 4 provides a detailed methodology to arrive at hazard rate estimates when mileage accumulation rates in the vehicle population of interest are available in the case of hard failures. It illustrates the methodology with an application example. Chapter 5 explains a methodology to estimate hazard function when mileage accumulation rate for the vehicle population is not available. Chapter 6 addresses the issue of bias in warranty data due to customer-rush near the warranty expiration limit. It treats soft failures as left censored and provides a methodology to arrive at nonparametric hazard rate estimates when information on mileage accumulation rates in the vehicle population is available. Chapter 7 deals with the bias in warranty data for the case when information on mileage accumulation rates in the vehicle population is not available.

Section III of the book consists of Chapters 8 and 9. Chapter 8 provides a methodology to assess the impact of changes in warranty period. It presents a simple method to assess the impact of new time/mileage warranty limits on the number and cost of warranty claims for components. It also discusses the bias in warranty cost estimates that may result in using cumulative cost per repair information and recommends a solution to the problem. Chapter 9 discusses the use of neural networks to forecast warranty performance in the presence of warranty growth phenomena. It compares the forecasting performance of log–log plot, radial basis function neural network, and multilayer perceptron neural network with application examples.

BIBLIOGRAPHIC NOTES

Suzuki (1985a) presented nonparametric estimation of lifetime distribution using generalized maximum likelihood estimator and discussed their statistical properties. Suzuki (1985b) examined the efficiency of a consistent estimator relative to maximum likelihood estimator when partial record of nonfailures can be obtained along with a complete record of failure. Nelson (1988, 1990, 1995) gave simple graphical methods for analyzing repairable systems data based on number and costs of repair and also provided confidence limits for the same. Kalbfleisch and Lawless (1988) showed that although failure time (for first failures) distributions can be estimated from failure data alone, by supplementing it with information on unfailed items more precise estimates could be obtained. They also provided some examples of truncated data sets and nonparametric methods for their analysis (Kalbfleisch and Lawless 1992). Lawless et al. (1995) proposed methods for fitting models based on warranty data and supplementary information about mileage accumulation. Hu and Lawless (1996) developed nonparametric estimation of means and rates of occurrences from truncated recurrent event data. Hu et al. (1998) considered nonparametric estimation for situations where censoring times on unfailed units are missing using maximum likelihood estimation and simple moment estimator.

2 Characterization of Warranty Data

> Every man has three characters: that which he shows, that which he has, and that which he thinks he has.

> —**Alphonse Karr**

OBJECTIVES

This chapter explains the following:

- Key points in the life cycle of a vehicle
- Important characteristics of automobile warranty data
- Warranty claims process
- Causes of unclean warranty data
- Different types of incompleteness associated with warranty data
- Difference between censoring and truncation

OVERVIEW

This chapter explains the key characteristics of automobile warranty data along with certain important definitions. Section 2.1 takes a look at certain key points in the life cycle of a vehicle. Section 2.2 provides steps involved in a warranty claim process. Section 2.3 details certain striking features of the warranty data. Section 2.4 then provides the causes that lead to unclean warranty data and also defines different types of incompleteness that warranty data exhibit.

2.1 LIFE CYCLE OF A VEHICLE

Similar to people, data too have characteristics. To identify the true character of a person, it is important that we spend some time with them and make an effort to understand them. The same is true for warranty data, too. To gain better understanding of the nature of information contained in warranty claims and its suitability for proper modeling and analysis, let us look at some of the key points in the life cycle of a vehicle as shown in Figure 2.1.

The entire life cycle of the vehicle can be divided into three major phases:

1. Phase A: Presale period
2. Phase B: Warranty period
3. Phase C: After warranty period

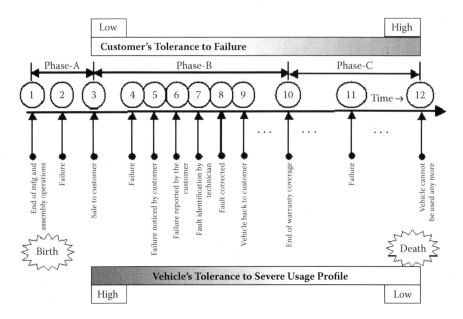

FIGURE 2.1   Key points in the life cycle of a vehicle.

The end of manufacturing and assembly operations at point 1 marks the birth of a vehicle. The design and manufacturing quality at this point is locked within the vehicle. From this point onward, customer usage and environmental conditions have major impacts over its life cycle. Failures occurring at point 2 can be largely due to errors associated with manufacturing and assembly operations. For illustration, consider the transmission assembly that transmits power from the engine to the drive wheels. Transmission fluid helps it deliver various functions by transferring power, cooling, lubricating, and cleaning. Severe damage of sealing material during assembly operations may lead to immediate transmission fluid leaks. To a large extent, such failures can be noticed at the assembly plant or at the dealer end. However, there is a small probability of such failures escaping to Phase B.

Sale of vehicle to the customer at point 3 starts the crucial phase B in the life cycle of vehicle. Warranty coverage provided by the automobile manufacturer in North America extends typically to 36K mi or 36 months from the date of sale, whichever occurred first. This is the phase in which a vehicle undergoes an acid test and the manufacturers' promises and customers' expectations are kept or broken. During this phase, customer-initiated vehicle repairs are carried out by an authorized service or repair center. Usually, there will be some gap between occurrence of failure at point 4 and the failure getting noticed by the customer at point 5. This gap will be minimum when the failure severely affects the functionality of the vehicle. When the severity is low, the gap between the occurrence of failure and the customer noticing it is expected to be high with a possibility of failure even going unnoticed. For example, if the transmission fluid leaks at a rate of about a drop every 24 h of driving the vehicle, the probability of such an event being noticed by the customer is low.

Yet again, depending on the severity of the failure, a gap exists between failure being noticed at point 5 and its being reported at point 6. For example, after noticing a drop of transmission fluid on the garage floor, the owner may choose to report it at the time of oil change. Thus, looking at the possibility of points 4, 5, and 6 being spread out over time, failure data arising from warranty claims in reality are very often left censored data. That is, we only know that the actual failure occurred at some point of time before the customer reported it, but exact time or mileage at failure remains unknown. However, to reduce complexity of analysis, very often it can be assumed that the gap between occurrence of failure and its reporting is negligible.

The duration of activities at points 7, 8, and 9 that ends with return of the vehicle to the customer after fault identification and correction largely depends on the capability of the repairing process. It is important to note that repeat warranty claims for the same failure mode between point 9 and point 10 can be a function of repair quality apart from design and manufacturing quality. Hence, to keep manufacturer quality separate from the repair quality, use of time or mileage to first warranty claim is recommended (Majeske et al. 1997; Hu et al. 1998). Phase B ends with the end of warranty coverage at point 10.

Failures at point 11 in Phase C can provide very useful feedback to the manufacturer for overall improvement of the vehicle. However, such information about failures is mostly restricted to the customer. Not many customers prefer authorized repair centers due to high cost of repair. Hence, failures in this phase of the life cycle are difficult to capture. Failure at point 12 renders the vehicle inoperative or where it cannot be profitably repaired, leading to the end of the life cycle of the vehicle.

The top horizontal bar in Figure 2.1 suggests that customers show lower tolerance to the occurrence of failures at the beginning of the vehicle life cycle than the later part. For example, a customer is very concerned about a minor transmission leak occurring in his or her 1-month-old car, but may not be as concerned about similar problem when the car is 10 years old. Goel and Singh (1999) nicely capture the concept of changing customer tolerance through a mathematical model. The bottom horizontal bar indicates that the vehicle tolerance to severe real-world usage conditions is high at the beginning of the life cycle and gradually decreases as the vehicle ages. Toward the end of the life cycle, many components are more susceptible to failure due to wearing out and aging.

Hence, failure information emanating from warranty claims is the result of a host of influencing factors during the entire life cycle of a vehicle. To extract "knowledge" from the unclean "information" contained in warranty claims, at times some kind of data refining becomes necessary. For an enlightening discussion on "information is not knowledge," see Deming (2000).

2.2   AN OVERVIEW OF THE WARRANTY CLAIM PROCESS

Figure 2.2 gives an overview of a general warranty claim process. The process starts with the user detecting the failure. If the vehicle is outside the warranty coverage, it is usually not taken to an authorized dealership owing to high cost of repair. For vehicles within the warranty coverage, the repair technician doing the repair work fills out the warranty claim sheet. The repair technician works to repair the vehicle

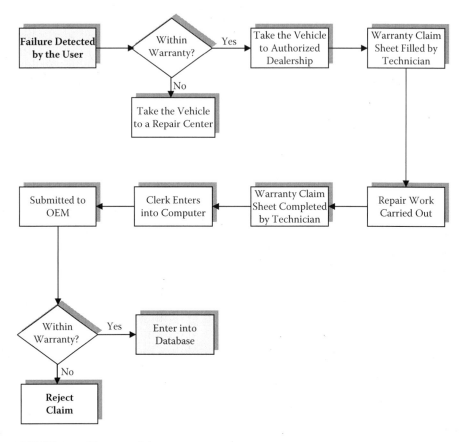

FIGURE 2.2   Warranty claim process.

and completes the warranty claim sheet once done. A clerk then enters the data into a computer, which is then submitted to the concerned original equipment manufacturer (OEM). The OEM checks whether or not the claim is valid and accordingly either enters it into the database or rejects it if it is found to be invalid.

2.3   AUTOMOBILE WARRANTY DATA: KEY CHARACTERISTICS

The primary reason for the existence of warranty data is known to be financial. As such, when a vehicle is repaired under warranty, information or data in excess of 50 different kinds are recorded in the warranty database. Among such data, two important variables related to a given failure mode are (1) months in service (MIS) from the date of sale of vehicle, and (2) mileage on odometer at the time of failure reporting.

Consider a two-dimensional warranty coverage and let M_1 and M_2 denote automobile warranty limits for mileage and time, respectively. The availability of failure data is restricted to vehicles with warranty claims, as depicted in Figure 2.3.

To enable extraction of certain key characteristics of warranty data, Figure 2.4 shows a plot of MIS versus mileage data for the first claims for a specific component-

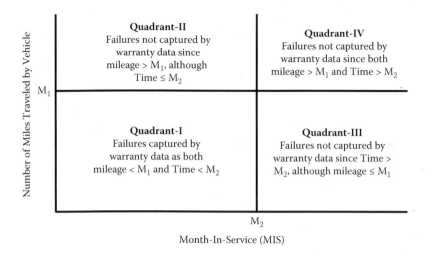

FIGURE 2.3   Limited availability of failure data from warranty claims.

level failure mode of a given model-year vehicle. The plot also gives a histogram of frequency of vehicles in 1K ("K" represents mileage in 000s) mileage band increments and histogram of frequency of vehicles that failed in each MIS. Because of the proprietary nature of the information, the specific vehicle details such as model year, part name causing the failure, vehicle line, etc., are not disclosed throughout this book.

The data for the scatter plot is based on first warranty claims involving 1002 vehicles. It excludes those data that represent repeat claims for the same vehicle and

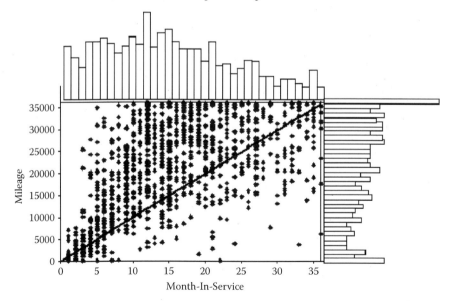

FIGURE 2.4   MIS versus mileage plot of first warranty claims.

also repairs done on the vehicle prior to delivery to the customer. Certain striking features about this plot are the following:

1. All the data points lay within 36 MIS and 36K mi of warranty coverage provided by the manufacturer. The data points spread out and take a rectangular shape. This is a typical pattern for any vehicle population covered by two-dimensional warranty. For certain vehicles that have a minimum of 5 year or 100K mi warranty coverage, the rectangular zone will be larger but again with no data available beyond 5 years and 100K mi. It is due to such incompleteness of data in estimating hazard rate that statistical treatment of the warranty data becomes necessary.

2. A diagonal line from (0, 0) to (36, 36K) for the rectangular warranty region divides it into an upper and lower triangle. It can be observed that the upper triangle contains a significantly higher number of data points than the lower triangle. Such a situation can arise under two situations. First, the mileage accumulation rate in the vehicle population is larger than the one represented by the diagonal line. This lead to more vehicles moving out of warranty mileage limit than the time limit. Second, there is a higher rate of mileage accumulation by the subpopulation of failed vehicles. If the mileage accumulation rate in the vehicle population is known, methods of modeling and analysis given in Suzuki (1985a), Kalbfleisch and Lawless (1988, 1992), and Lawless et al. (1995) are useful.

3. There is a heavy concentration of data points between (0, 0) and (5, 5K). Out of a total of 1002 vehicles with claims within the warranty period, there are 91 claims in this region alone. If claims were distributed uniformly over the entire rectangular warranty region, one would expect 0.8 claims ($1002/(36 \times 36) = 0.8$) per unit area. However up to 5-MIS and 5K-mi claims per unit area is observed to be 3.6 ($91/(5 \times 5) = 3.6$), which is more than four times the expected number. Similarly, the number of claims per unit area up to (1, 1K), (2, 2K), (3, 3K), and (4, 4K) are 22.0, 9.5, 6.4, and 4.7, respectively. As mentioned earlier, this could occur owing to a high number of manufacturing- or assembly-related defects being detected during the early part of the vehicle life cycle in addition to the usage-related failures (Majeske 2003).

4. There are a few vehicles that are quite close to the x-axis, which represents MIS. For example, a vehicle at (22, 1) with a mileage accumulation rate of $1/22 = 0.05$ mi/month has an unusually low rate. It is highly likely, considering the volume of warranty claims processed each day, that there is a typographical error while entering the mileage data into the warranty database. It is advisable to screen out such data before carrying out any statistical analysis.

5. The right side of Figure 2.4 shows "spikes" in first warranty claims, both at the beginning and toward the end of the warranty mileage limit. The spike at the beginning can be attributed to the existence of manufacturing or assembly defects in addition to the usage-related failures (Majeske 2003). The spike observed in the last mileage band of the histogram, that is, 35K–36K mi, could be attributed to the customer-rush near the warranty

limit (Rai and Singh 2004a). Such a phenomena is expected to occur, especially for soft failures, where vehicle users delay failure reporting until warranty is about to expire. Although actual failure for first claims affected by the customer-rush phenomena are expected to have occurred prior to the reported time, the actual MIS or mileage at the time of failure is not known. Such data are said to be *left censored*.

6. The top portion of Figure 2.4 also shows a histogram of the number of vehicles with first claims during MIS = 1 to 36. It can be observed that the number of first claims are relatively higher during MIS = 1 to 21, as compared to the number of first claims observed during MIS = 22 to 36. The lower number of claims during the later period can be attributed to vehicles leaving the population owing to exceeding the warranty mileage limit. To enable comparison of rates of occurrence of first failures at different MIS values, one needs to estimate hazard rates at each MIS.

2.4   TWO INHERENT CHARACTERISTICS OF WARRANTY DATA: UNCLEANLINESS AND INCOMPLETENESS

The six observations discussed in Section 2.3 point toward two inherent characteristics of automobile warranty data, which needs attention in field reliability studies. They are

a. Uncleanliness
b. Incompleteness

2.4.1   UNCLEAN WARRANTY DATA

Warranty data are known to be messy and unclean for reasons including data entry errors, incorrect binning of claims, and inaccurate reporting of failures (Iskandar and Blischke 2003; Pal and Murthy 2003; Sander et al. 2003; Suzuki et al. 2001; Robinson and McDonald 1990). Thus, it is always helpful to screen the warranty data before undertaking any detailed statistical analysis. Warranty claims with obvious errors may need to be eliminated from further analysis or corrected, if feasible. For example, the warranty data in Figure 2.4 includes a claim with MIS = 22 and mileage = 1 mi, which is likely to be a result of an unintended error while recording information. A sound strategy is required to address such data before proceeding further. Causes that lead an automobile warranty data to be unclean and messy can be broadly classified into three main categories as shown in Figure 2.5.

2.4.1.1   Type of Failure Mode

Customers may respond differently to different types of failure modes experienced during the warranty period. Vehicle failures experienced by users can be mainly classified into two types: hard failures and soft failures. Hard failures are those that make a vehicle inoperative or unusable until repaired. For example, "engine does

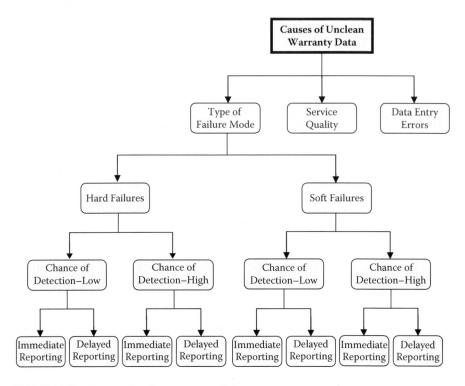

FIGURE 2.5   Causes of unclean warranty data.

not start" or "engine stops running" belongs to the hard failure category. Chance of detecting a hard failure may be high or low. For example, the failure of one of the rear brake lights may get detected only at the time of visit to the dealer for some other service or repair. Thus, time to the claim minus time to the failure in such cases could be greater than zero. Hard failures with a high chance of detection by users are generally immediately reported. Also, due to largely immediate reporting of such failures the time to claim minus time to failure is approximately zero, leading to a more accurate failure time or mileage data in warranty claims. However, it is to be noted that even hard failures with a high chance of detection by users may not be reported for claim immediately. For example, immediate reporting of an engine stops running failure may depend on whether or not the vehicle could be restarted after experiencing the failure.

Soft failures are those that result in degraded performance, but the vehicle can still be operated. Some examples of soft failure are minor oil leaks, engine slow to start, unusual engine noise, etc. For soft failures, the chance of detection by the user (low/high) and reporting time (immediate/ delayed) influence the pattern of reported failures in the warranty data. For soft failures, time to claim minus time to failure is likely to be greater than zero.

2.4.1.2 Service Quality

The process of fault identification and correction largely depends on the capability of the repairing process. Occasional misclassification of failure modes and misdiagnosis of the vehicle can lead warranty data to be unclean for reliability studies. Often, a repeat repair for the same failure mode is a function of repair quality apart from the design and manufacturing quality. Hence, in order to separate manufacturer quality from service quality, use of time or mileage to first warranty claim is recommended (Hu et al. 1998; Majeske et al. 1997).

2.4.1.3 Unintended Data Entry Errors

When a vehicle fails within the warranty period, data or information in excess of 50 different kinds are obtained. Although most of the data entry process is computerized, there are certain repair-related entries that the repair technician is required to carry out. Due to the human-intensive nature of data entry, unintended data entry errors could also lead to unclean warranty data.

2.4.2 Incompleteness of Warranty Data

Incompleteness of automobile warranty data can also be broadly classified into three categories, as shown in Figure 2.6.

2.4.2.1 Truncated Warranty Data

As failures outside warranty time and mileage are not part of warranty data, such data sets are said to be *right truncated* at the time and mileage limit, respectively. Consider a situation where we wish to use the warranty data shown in Figure 2.4 only for those first claims that occurred between 1K mi and 35K mi. In this case, the number of claims and mileage values for vehicles with first claims below 1K and above 35K mi is assumed to be unknown at each MIS value. The mileage distribution of vehicles with first claims at different MIS values in this case is said to be *left truncated* with truncation point at 1K mi and *right truncated* with truncation point at 35K mi. Data sets that are both left truncated and right truncated are said to be *doubly truncated* data sets. Data sets that are either right truncated or left truncated are said to be *singly truncated* data sets.

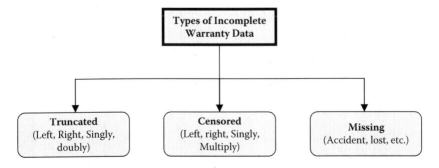

FIGURE 2.6   Three types of incomplete warranty data.

2.4.2.2 Censored Warranty Data

Consider three warranty claims with reported MIS and mileage values of (12, 35600), (25, 35930), and (32, 35995) with a warranty period of 36 MIS or 36K mi, whichever comes first. An engineering analysis of the condition of field-returned parts indicates that the actual failure for these claims occurred before the reported time or mileage. If we treat MIS as a life variable, then the three claims are said to be *left censored* at MIS = 12, MIS = 25, and MIS = 32, respectively. Also, because the three left-censored claims have different censoring MIS values, they are said to be a case of *multiply censored on the left*, as shown in Figure 2.7 (Nelson 1982).

Similarly, when mileage (in intervals of 1K mi) is considered as a life variable, the three left-censored claims are said to be *singly censored on the left* at 35K mi.

Consider another situation where there are three vehicles without any warranty claim and with MIS and mileage values of (20, 25K), (30, 32K), and (34, 27K). Since failure time or mileage for these three vehicles are known only to be beyond their present running time or mileage, the observed data are said to be *multiply censored on the right* for both MIS and mileage as life variable (Petkova et al. 2000).

As different statistical procedures apply to truncated and censored data sets, it is important to understand the differences between the two (Figure 2.8).

Figure 2.8a shows vehicles that failed before warranty mileage limit *M*. The cross sign represents failures within the warranty limit. Such data are called *right truncated*, meaning both the count and the number of miles on the odometer at the time of failure above the truncation mileage *M* are completely unknown. In contrast, Figure 2.8b shows the case of right-censored data. Suppose that a fleet comprising five vehicles is under study and three have failed before "C" miles. The remaining two vehicles, represented by the oval sign at the point C, survive. The data on unfailed vehicles are called *right censored*, where the number of observations above C miles is known, but the exact failure point is unknown.

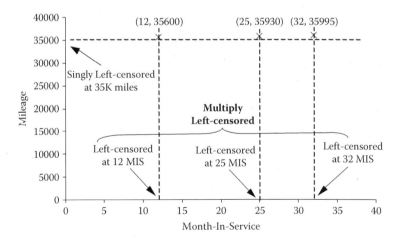

FIGURE 2.7 Example of singly and multiply left-censored data.

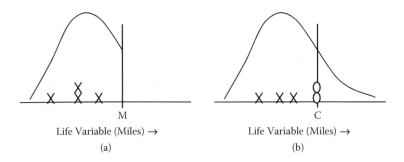

Life Variable (Miles) →   Life Variable (Miles) →

(a)   (b)

FIGURE 2.8   Difference between truncated and censored data sets: (a) right-truncated data, (b) right-censored data.

2.4.2.3   Missing Warranty Data

In the vehicle population of interest, some vehicles are lost owing to accident or theft, and as such, potential failures may not become part of the warranty data sets. Since, such missing warranty data are not expected to be large in number when compared to vehicle population, they are ignored for the purpose of reliability analysis.

2.5   SUMMARY

Key points from the chapter are summarized in the following:

* Failure information emanating from warranty claims is the result of a host of influencing factors during the entire life cycle of a vehicle. To extract "knowledge" from the unclean "information" contained in warranty claims, statistical treatment of the data is necessary.
* Causes of unclean nature of warranty data can be classified into three categories: type of failure mode experienced by the user, service or repair quality, and unintended data entry errors. Failure modes can be mainly of two types: hard failure and soft failure. Reported time and mileage values captured by warranty data are likely to be more accurate for hard failures than for soft failures.
* Incompleteness of warranty data can be broadly classified into three types: truncated, censored, and missing. Hard failures result in truncated warranty data set for the failed units and right-censored data for unfailed units, if available. Soft failures result mainly in left-censored warranty data for the reported failures in warranty data and also right-censored data for unfailed units, if available.

Section II

Strategies and Methods
for Reliability Analysis
with Warranty Data

3 Strategies for Reliability Analysis from Warranty Data

If you always do what you've always done, you'll always be where you are now.

—Anonymous

OBJECTIVES

This chapter explains the following:

- Different levels of analysis, from customer-reported concerns (CRCs) to root cause analysis for reliability and robustness improvement problems
- Approaches for adjustments or modifications to the hazard function when mileage accumulation rates in the vehicle population are either known or not known
- Example of estimating hazard rate by adjusting the risk set

OVERVIEW

In Chapter 2, key characteristics of the warranty data were discussed. In this chapter, we discuss the strategies to arrive at appropriate hazard rate estimates in light of such characteristics. Section 3.1 provides a three-level process for analyzing warranty data for prioritization and directions for reliability and robustness improvement problems. Section 3.2 gives strategies to modify the hazard function in the presence or absence of mileage accumulation rates in the vehicle population.

3.1 FROM CUSTOMER CONCERNS TO ROOT CAUSES

Warranty data are stratified in several ways to prioritize and get directions for reliability and robustness improvements. Concerns reported by a customer and the part reported to be the cause of the problem by the repair technician are two such important factors. Both these factors have one-to-many relationships with each other as shown in Figure 3.1.

Figure 3.1 shows that a concern reported by several customers could have multiple parts reported to be the cause and vice versa. Although it may be difficult to generalize whether to initiate reliability and robustness improvement studies with customer concerns or with parts reported to be causing the concern, some comments can be

35

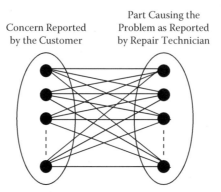

FIGURE 3.1   One-to-many relationships between the concerns reported by customers and the part causing the problem as reported by the repair technician.

made about their differences. The most important distinction is in terms of misclassification rates. Because the concerns reported by customers can be easily verified at the time of claim, misclassification errors are expected to be low. For example, engine does not start, CD player problems, or oil leaks have virtually remote chances of being classified as some other concern, because they can be easily verified at the time of failure reporting by the customer. However, identifying the right part causing the failure can very often depend on the technical capability of the repair technician. This, in turn, sometimes can lead to a relatively high chance of misclassification and misdiagnosis. On the other hand, for an engineer working on a reliability and robustness improvement problem, it is easier to think in terms of components or subsystems, as any design, manufacturing, or service fix will eventually impact a component.

A sound strategy to identify and prioritize areas for reliability and robustness improvement as well as warranty cost reduction would broadly involve three levels of analysis, as shown in Figure 3.2 with an example. The Pareto chart in Figure 3.2 for level-1 analysis identifies the top CRCs. The example shows concerns C1 to C6 with shaded rectangles indicating the top six concerns (reported by customers) that contribute 80% or more to the overall warranty claims. Hazard plots for the top concerns C1, C2, etc., indicate whether or not the problem is getting worse with time. For example, an increasing failure rate pattern for C2 obtained from the hazard plot may be more cause for concern than a decreasing failure pattern for C1, even though the number of claims for C1 is more than that for C2 at the time of analysis. In addition, as discussed in Chapter 1, the pattern of failure rates obtained from the hazard plots would also indicate the type of noise factors that influence the field failures. Warranty data stratified according to concerns reported by the customers and the associated hazard rate estimates are expected to be fairly accurate because of minimal chance of misclassification.

Each of the top customer concerns can then be broken down further at level-2 analysis into parts reported to be causing a particular CRC. In Figure 3.2, a Pareto chart shows the top five parts, P1 to P5, that account for greater than 80% of the claims associated with the CRC C1. A cautious approach is required when estimating hazard rate at level 2, as the chances of misclassification or misdiagnosis by the

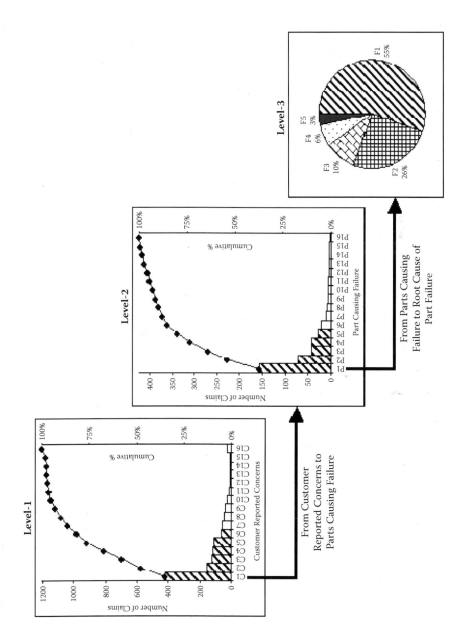

FIGURE 3.2 From customer-reported concern (CRC) to detailed analysis of part causing failure to identify the root causes.

repair technician are comparatively higher. Feedback from level-3 analysis can help to obtain more accurate estimates of hazard rates at level 2. The analysis at level 3 aims to obtain cause of the failure for the parts reported to be responsible for customer concerns. The following methods can be used for such an analysis:

- Reading the repair technician's comments in the warranty claim records
- Talking to the repair technicians for clarification of comments in the claims that do not clearly indicate cause of the part failure
- Performing engineering analysis of parts returned from the dealers
- Checking the claims history of the vehicle

The pie chart example in Figure 3.2 at level-3 analysis indicates that failure causes F1 and F2 together result in about 81% of the part P1 failures. The analysis at this level helps to identify warranty claims that are misclassified, that is, the part causing the failure is other than that reported to be the cause. It also helps to identify parts for which no problems were found and were incorrectly diagnosed to be causing the customer concern. The claims for which no problems have been found with the parts reported to be causing the customer concern should be excluded when estimating the hazard rates for the parts. Similarly, the misclassified claims should be included with appropriate parts at the time of hazard rate estimation at level 2.

It is to be noted that when going from level 1 to level 3, the analysis at times can become too involved and time consuming as the number of claims increases. However, the advantages of such an analysis would always more than compensate for the time and effort. The benefits include feedback on inclusion of realistic noise factors for design verification and testing of future model-year vehicles that would make products more reliable and robust to varying field conditions and thus help to reduce the warranty cost. Once proper component- or subsystem-level fixes are in place, the Pareto charts and hazard plots based on the new data can be used to assess the effectiveness of the design, manufacturing, or service fixes.

3.2   STRATEGIES FOR HAZARD FUNCTION ESTIMATION

According to Kaplan and Meier (1958), nonparametric estimation procedure is identified by the following condition: the class of admissible distributions from which the best-fitting distribution is chosen is the class of all distributions. Meeker and Escobar (1998) suggest that in reliability studies the data analysis should begin with analytical and graphical tools that do not require strong model assumptions. In this book, we focus on nonparametric estimates of the hazard function.

Different types of incompleteness of warranty data sets discussed in Chapter 2 call for different approaches to arrive at the estimates of the hazard function. Let N be the total number of vehicles in the field. Let n_t^* denote the number of first warranty claims at months in service (MIS) = t, and $N(t)$ denote the number of vehicles in the field without any claims at the beginning of MIS = t. Also, let V_t denote the total number of vehicles in the field up to MIS = t. Assuming an unlimited warranty mileage limit for the vehicle population under study and a warranty time limit of M_2, an estimate of hazard function may be obtained as follows:

$$h_1(t) = \frac{n_t^*}{N(t)} \ , t = 1, 2, \ldots, M_2 \text{ and } M_1 = \infty \qquad (3.1)$$

where $N(t) = V_t - \sum_{j=1}^{t-1} n_j^*$ for $t = 2, 3, \ldots, M_2$ and $N(1) = V_1 = N$.

EXAMPLE 3.1

An example in Table 3.1, based on hypothetical data, uses Equation 3.1, assuming no limit on the warranty mileage.

TABLE 3.1

Hazard Rate Estimates without Mileage Warranty Limit

MIS (t)	n_t^*	V_t	$N(t)$	$h(t)$	$H(t)$
1	141	200,000	200,000	0.00071	0.00071
2	103	200,000	199,859	0.00052	0.00122
3	117	200,000	199,756	0.00059	0.00181
4	102	200,000	199,639	0.00051	0.00232
5	84	200,000	199,537	0.00042	0.00274
6	95	200,000	199,453	0.00048	0.00321
7	101	200,000	199,358	0.00051	0.00372
8	79	200,000	199,257	0.00040	0.00412
9	99	200,000	199,178	0.00050	0.00461
10	95	200,000	199,079	0.00048	0.00509
11	88	200,000	198,984	0.00044	0.00553
12	111	200,000	198,896	0.00056	0.00609
13	109	200,000	198,785	0.00055	0.00664
14	109	200,000	198,676	0.00055	0.00719
15	100	200,000	198,567	0.00050	0.00769
16	96	200,000	198,467	0.00048	0.00818
17	92	200,000	198,371	0.00046	0.00864
18	115	200,000	198,279	0.00058	0.00922
19	93	200,000	198,164	0.00047	0.00969
20	76	200,000	198,071	0.00038	0.01007
21	81	200,000	197,995	0.00041	0.01048
22	75	200,000	197,914	0.00038	0.01086
23	53	200,000	197,839	0.00027	0.01113
24	53	200,000	197,786	0.00027	0.01140
25	64	200,000	197,733	0.00032	0.01172
26	38	180,000	177,669	0.00021	0.01193
27	23	160,000	157,631	0.00015	0.01208
28	22	140,000	137,608	0.00016	0.01224

In Table 3.1, $N(1) = V_1 = N = 200{,}000$, which is the total number of vehicles in the field. Column $N(t)$ for MIS = 2 onward is obtained by subtracting total first claims up to previous MIS from V_t, and it gives the number of vehicles at risk of failure at the beginning of each MIS. $V_{26} = 180{,}000$ indicates that 20,000 vehicles out of the total population have reached 25 MIS, but are yet to reach 26 MIS. In this example, the calculations are based on occurrence of 0 claims from these 20,000 vehicles, resulting in $N(26) = 180{,}000 - 2{,}331 = 177{,}669$. However, if there were, say, 10 first claims out of a total of 2,331 at MIS = 25 from the 20,000 vehicles yet to reach 26 MIS, $N(26)$ would have been 177,679, indicating the number of vehicles at risk of failure at the beginning of MIS = 26. This is because there would be 10 additional vehicles in $V_{26} = 180{,}000$ that are yet to fail at the beginning of 26 MIS. Figure 3.3 shows the hazard plots for the data in Table 3.1.

It can be seen from Figure 3.3a that the hazard rate is largely constant with fluctuations around 0.0005 until about MIS = 20 and then shows a decreasing trend. The cumulative hazard plot for the same data in Figure 3.3b shows a smoother curve with constant failure rate indicated by a straight line up to about MIS = 20 and the decreasing failure rate indicated by the curve bending downward.

In the presence of a finite warranty mileage limit M_1, Equation 3.1 requires a modification. Either the denominator or numerator, respectively, may be modified, depending on whether or not the mileage accumulation rates in the vehicle population are known. When the warranty mileage limit M_1 is finite, a failure may occur within or outside the warranty mileage limit of M_1, whereas the claim would always be within M_1. When the mileage accumulation rate in the vehicle population is known, the denominator can be modified to represent the number of vehicles at risk of first claim within the warranty mileage limit M_1. When the mileage accumulation rate in the vehicle population is not known, the numerator of Equation 3.1 can be

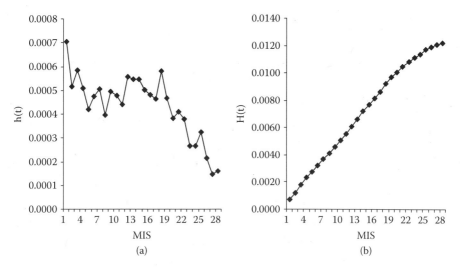

FIGURE 3.3 Hazard plots for the data in Table 3.1: (a) hazard plot, (b) cumulative hazard plot.

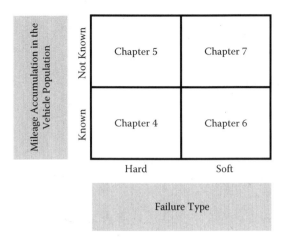

FIGURE 3.4 Chapters based on type of failure and knowledge about mileage accumulation in the vehicle population.

modified by estimating the total number of first failures at MIS = t by estimating the first failures within and outside the mileage limit M_1.

For accurate estimation of the hazard function, the type of failure, that is, hard or soft failures, also needs to be taken into account. A soft failure with a low chance of detection by the user may gradually aggravate with usage until it gets detected. Once detected, it may be reported immediately or at a convenient date by the user. The MIS and mileage values reported in warranty claims for soft failures with a low chance of detection by user should thus be treated as left censored. In other words, the actual time or mileage at failures should be treated as occurring before the reported time or mileage. Claims associated with soft failures having a high chance of detection by the user may or may not be left censored. The methodologies to deal with situations when mileage accumulation rates in the vehicle population are either known or unknown and the type of failure is hard or soft is given in different chapters, as shown in Figure 3.4.

3.3 SUMMARY

Key points from the chapter are summarized as follows:

- For reliability and robustness improvement problems using warranty data, a three-level analysis process is suggested. The level-1 analysis involves Pareto analysis of CRCs to identify top concerns and hazard plots of the top CRCs. Level-2 analysis involves Pareto analysis of the parts for each of the top CRCs. To arrive at accurate estimates of the hazard rates for each part, results of failure cause analysis at level 3 are used. Claims with misdiagnosed parts having no troubles found are excluded when estimating the hazard rate. The claims found to be misclassified are included with appropriate parts before arriving at hazard rate estimates.

- The numerator or denominator of the hazard function is modified depending on whether or not the mileage accumulation rates in the vehicle population are known. When the mileage accumulation rates in the vehicle population are known, the denominator of the hazard function is modified to obtain vehicles at risk of first claims. On the other hand, when the mileage accumulation rates in the vehicle population are not known, the numerator of the hazard function is modified to obtain the total number of first failures within and outside the warranty mileage limit. In addition, the failure time or mileage for soft failures is treated as left censored when arriving at estimates for the hazard function.

4 Hard Failures with Known Mileage Accumulation Rates

Everything should be made as simple as possible, but not simpler.

—**Albert Einstein**

OBJECTIVES

This chapter explains the following:

- Hazard function estimation for hard failures with known mileage accumulation rates
- Method for mileage accumulation rate modeling
- A four-step methodology with example

OVERVIEW

This chapter shows how mileage accumulation rate information can be used to arrive at hazard function estimates for hard failures. Section 4.1 provides the hazard function with the risk set adjusted for vehicles that move out of the warranty mileage limit. Section 4.2 gives a method for modeling mileage accumulation rates in the vehicle population. Section 4.3 describes a four-step methodology to estimate the hazard rates. Section 4.4 illustrates the use of the four-step methodology with an application example.

4.1 RISK SET ADJUSTMENT IN HAZARD FUNCTION

As discussed in previous chapters, the reported months in service (MIS) or mileage values for the claims associated with hard failures are likely to be more accurate than the similar value for soft failures. Accordingly, in this chapter, it is assumed that the time and mileage in warranty claims are the same as the time and mileage at failure. In the presence of the warranty mileage limit, say M_1, the information about the numerator of Equation 3.1 (Chapter 3) is available only for the first claims within warranty coverage. To use the available information, the denominator in the equation needs to be adjusted for vehicles that have crossed the warranty mileage limit of M_1 miles. Let W_t ($t = 1, 2, 3, \ldots, M_2$) be a random variable denoting miles driven by a vehicle at MIS = t. The number of vehicles outside the warranty mileage limit M_1

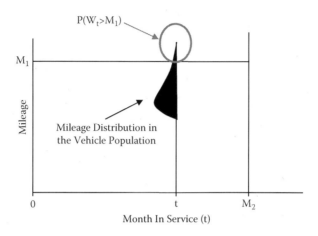

FIGURE 4.1   Risk set adjustment using mileage distribution in vehicle population.

at MIS $= t$ can be obtained as $N(t) \times P[W_t > M_1]$. Equation 3.1 (Chapter 3) can then be rewritten as follows:

$$h_2(t) = \frac{n_t^*}{N(t) - N(t)P[W_t > M_1]} = \frac{n_t^*}{N(t)P[W_t \leq M_1]} \qquad (4.1)$$

Figure 4.1 shows the distribution of mileage in a population of vehicles at MIS $= t$. $P[W_t > M_1]$ is the area under the curve beyond the warranty mileage limit M_1. This helps to obtain the number of vehicles at risk of first claims within the warranty coverage.

4.2   MODELING OF MILEAGE ACCUMULATION RATE IN THE VEHICLE POPULATION

A warranty database usually contains mileage accumulation data for only those vehicles that fail within the warranty period. However, estimation of parameters for mileage accumulation requires information on MIS and mileage for unfailed vehicles too. Let N be the total number of vehicles in the field. Let w_i ($i = 1, 2, ..., n$; $n \leq N$) in miles per month denote the mileage accumulation rate of the i-th vehicle. Let $g(w)$ and $G(w)$ be the probability density function (pdf) and distribution function (df), respectively, for random variable w. The mileage accumulation rate w_i for the i-th vehicle is obtained as follows:

$$w_i = \frac{u_i}{t_i}, \ i = 1, 2, ..., n \qquad (4.2)$$

where u_i is the mileage of the i-th vehicle, and t_i is the corresponding MIS. For example, a vehicle that has traveled 12K mi at the time of MIS $= 10$ will have a mileage accumulation rate of $12K/10 = 1200$ mi/month. Although the vehicle is not expected to accumulate mileage at the rate of 1200 mi/month at every point in its life cycle, it is reasonable to assume a linear trend between mileage at MIS $= t$ versus t.

The parameters of $g(w)$ can be estimated by taking a random sample from the population of N vehicles. A sample from only failed or only unfailed vehicles may be biased if the mileage accumulation rate is correlated with the failure mode under study. Thus, a random sample would ensure that both failed and unfailed vehicles form part of the sample. Although follow-up surveys can be used for estimating the parameters of $g(w)$, they can often be very costly. Sometimes, the manufacturer recalls a large population of vehicles owing to quality and reliability related issues. Such data can also be utilized to estimate parameters of mileage accumulation distribution. Note that use of follow-up data or recall data need not be restricted to the model year under study, as the mileage accumulation rate for a given type of vehicle is expected to be independent of the model year.

Consider a case in which the lognormal distribution provides a good fit to the mileage accumulation rate data. The pdf of a lognormal distribution is given by

$$g(w) = \frac{1}{\sqrt{2\pi}\sigma w} e^{-\frac{1}{2}\left(\frac{\log(w)-\mu}{\sigma}\right)^2}, \quad w > 0 \qquad (4.3)$$

For $W \sim LN(\mu, \sigma^2)$, μ is called the *location parameter* and σ is called the *scale parameter*. The mean and variance of a lognormal distribution are respectively given by $e^{\mu+\frac{1}{2}\sigma^2}$ and $e^{2\mu+\sigma^2}\left[e^{\sigma^2}-1\right]$. Assuming the mean mileage accumulation rate to increase linearly with MIS with constant σ, the distribution of miles on odometer at MIS $= t$ is given by

$$W_t \sim LN[\mu + \ln(t), \sigma^2], \quad t = 1, 2, \dots, M_2 \qquad (4.4)$$

Figure 4.2 gives pdf $g(w_t)$ and cumulative density function (cdf) $G(w_t)$ at MIS $= 2, 5,$ and 10 with $\hat{\mu} = 7.279 + \ln(t)$ and $\hat{\sigma} = 0.688$. These estimates for the parameters of the mileage accumulation rate distribution translate to approximately 1450 mi/month or 17400 mi/year.

Figure 4.2a shows how the probabilities of a vehicle with certain mileage change with increasing MIS. For example, at MIS $= 2$, the probability of a vehicle accumulating about 3000 mi is much higher when compared to the probability of its accumulating 36,000 mi, which is negligible. Similarly, Figure 4.2b helps to obtain the proportion of vehicles in the population within or outside the warranty mileage limit. It can be seen that at MIS $= 2$, all the vehicles in the population are expected to have mileages below 36,000 mi. On the other hand, at MIS $= 10$, approximately 10% of the vehicles in the population are expected to have mileage more than 36,000 mi, or approximately 90% of the vehicles are expected to have less than 36,000 mi.

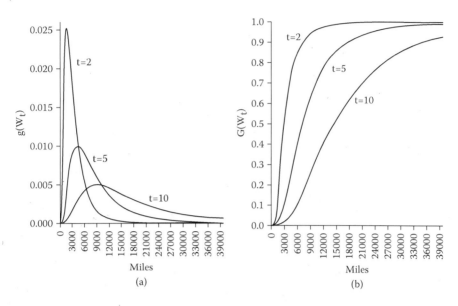

FIGURE 4.2 Mileage accumulation rates at MIS = 2, 5, and 10 for Wt ~ LN[7.279 + ln(t), 0.688²]: (a) probability density function g(w₁) and (b) cumulative density function G(w₁).

EXAMPLE 4.1

Consider mileage accumulation rate in the vehicle population under study to be such that $W \sim LN(7.279, 0.688^2)$. Also, at 15 MIS, $n^*_{15} = 46$, $N(15) = 120,000$, and $M_1 = $ 36K mi. Then, using $W_{15} \sim LN[7.279 + \ln(15), 0.688^2]$, we get $P[W_{15} < 36K] = 0.768$. For obtaining such probabilities, the NORMDIST function (based on normal distribution) available in Excel software can be used. This function makes use of the fact that when W follows the lognormal distribution, $\log(W)$ follows the normal distribution. In this example, NORMDIST($LN(36000)$, 7.279 + $LN(15)$, 0.688, TRUE) gives 0.768. Thus, the number of vehicles at risk of resulting in first claim within the warranty mileage limit is $0.768 \times 12,000 = 92,160$. Thus, using Equation 4.1 we get $h_2(15) = 0.000499$.

Estimates of mileage accumulation rates in the vehicle population also provide design engineers with feedback on usage profiles of typical 90th or 95th percentile customers. Such information helps in the design of products that meet the durability requirements of 90% or 95% customers for the useful life of the vehicle targeted by the design. For example, from the lognormal parameters $\hat{\mu} = 7.279 + \ln(t)$ and $\hat{\sigma} = $ 0.688, we find that a 90th (or 95th) percentile vehicle would accumulate mileage at the rate of 3500 mi (or 4495 mi). Thus, if the company plans to design a subsystem or component that meets the durability requirement of 90% or 95% of its customers for a useful life of 7 years, then the durability targets could be accordingly set for at least 294,000 mi or 337,580 mi, respectively.

Once mileage accumulation rate in the vehicle population is estimated, arriving at the hazard rate using Equation 4.1 does not pose much of a problem.

4.3 A FOUR-STEP METHODOLOGY

A four-step methodology to arrive at the hazard rate estimate given by Equation 4.1 is shown in Figure 4.3.

Step 1: For every vehicle, warranty claims with minimum mileage are retained and repeat claims are screened out. The purpose is to keep manufacturing quality separate from service or repair quality. Misclassified and misdiagnosed claims are identified through one or more of the methods discussed in Chapter 3 and excluded from further analysis.

Step 2: This step deals with the issue of incompleteness of warranty data. Using the MIS and mileage values available for unfailed vehicles, estimates of population parameters for the mileage accumulation rates are obtained.

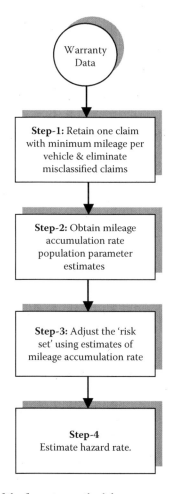

FIGURE 4.3   Flowchart of the four-step methodology.

Step 3: Using the mileage accumulation rate distribution, the denominator of the hazard function is adjusted to obtain the risk set, that is, the number of vehicles that have not failed yet and are within the warranty mileage limit.

Step 4: The estimate of the hazard rate at each MIS is obtained using Equation 4.1.

4.4   AN APPLICATION EXAMPLE

Step 1: For a component-level failure mode of a vehicle, 2574 claims were reported during MIS = 1 to 28. After completing the first step, 160 repeat claims were discarded. Also, using the information in the technician comments and returned parts analysis, 14 claims were found to be misclassified and were therefore excluded. The remaining 2400 claims were considered for further analysis.

Step 2: Recall that data from a previous model-year vehicle was used to obtain the mileage accumulation rate, as no such data was available from the current model-year vehicle. Figure 4.4 shows four-way probability plots of the mileage accumulation rates.

It can be visually observed from Figure 4.4 that the lognormal distribution provides a good fit for the mileage accumulation data. Based on the lognormal distribution, an Anderson–Darling (AD) statistic value of 0.52 is obtained with significance level p of 0.19. AD test is a goodness-of-fit test based on empirical distribution function. It measures the distance or discrepancy between the empirical and the specified (or assumed) distributions (Kececioglu 1993). A value of significance level p close to 1 indicates

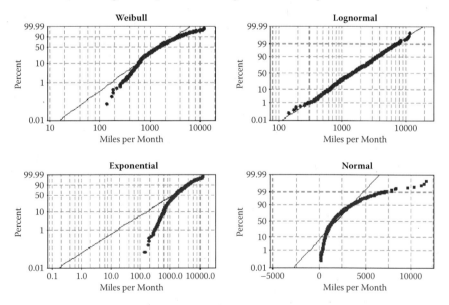

FIGURE 4.4   Four-way probability plots of mileage accumulation rates.

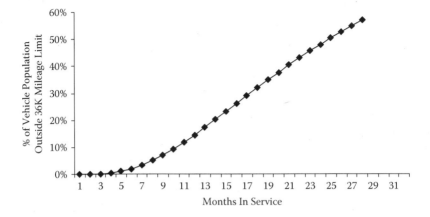

FIGURE 4.5   Percentage of vehicle population outside the 36K warranty mileage limit.

closeness of the data set to the chosen distribution and vice versa. The estimates of location and shape parameter of the lognormal distribution are obtained as 7.28 and 0.689, respectively. These estimates also indicate approximate mean mileage accumulation of 1840 mi/month and standard deviation of 1437 mi/month.

Step 3: Using the mileage accumulation rates in the vehicle population, the denominator in the hazard function was adjusted for vehicles crossing the warranty mileage limit. The percentage of vehicle population outside the 36K mileage limit at different MIS values is shown in Figure 4.5.

Step 4: The hazard rate estimates obtained using Equation 4.1 are given in Table 4.1, and the corresponding hazard plot is shown in Figure 4.6. We observe from the figure that the cumulative hazard plot mainly shows a constant failure rate (CFR) pattern, indicating that the failure mode is likely a result of a random failure mechanism and is influenced by N3, N4, and N5 categories of noise factors. The absence of any major pattern depicting decreasing or increasing failure rates implies that looking into N1 and N2 categories of noise factors may not help much in preventing the occurrence of this failure mode.

4.5   SUMMARY

The data quality in the warranty data can be effectively addressed by using different levels of screening. Retaining claims with minimum reported mileage per vehicle helps to separate manufacturer quality from repair quality. Eliminating the misclassified and misdiagnosed claims further cleans the data set and prepares it for proper reliability analysis. Using the mileage accumulation rates in the vehicle population, the proportion of vehicle population outside the warranty mileage limit is determined to arrive at risk of first claims at any MIS. Hazard rate estimates are then arrived at using Equation 4.1.

TABLE 4.1

Hazard Rate Estimates for Hard Failures with Known Mileage Accumulation Rates in the Vehicle Population and Warranty Mileage Limit of 36,000 mi.

MIS t	Total Claims	First claims	Claims Misclassified or Misdiagnosed	$n_t^{'}$	Number of Vehicles in Field	$N(t)$	$P[W_t > M_1]$	$N(t) \times P[W_t \geq M_1]$	$h(t)$	$H(t)$
1	143	141	—	141	200,000	200000	0.000002	199999.7	0.0007	0.0007
2	105	103	—	103	200,000	199859	0.000125	199833.9	0.0005	0.0012
3	126	117	—	117	200,000	199756	0.001063	199543.7	0.0006	0.0018
4	108	102	—	102	200,000	199639	0.003977	198845.1	0.0005	0.0023
5	91	84	—	84	200,000	199537	0.009911	197559.4	0.0004	0.0027
6	104	95	—	95	200,000	199453	0.019476	195568.5	0.0005	0.0032
7	110	101	—	101	200,000	199358	0.032837	192811.7	0.0005	0.0038
8	87	79	—	79	200,000	199257	0.049824	189329.2	0.0004	0.0042
9	110	99	—	99	200,000	199178	0.070058	185224.1	0.0005	0.0047
10	102	95	—	95	200,000	199079	0.093047	180555.3	0.0005	0.0052
11	94	88	—	88	200,000	198984	0.118268	175450.6	0.0005	0.0057
12	120	111	—	111	200,000	198896	0.145206	170015.2	0.0007	0.0064
13	115	109	—	109	200,000	198785	0.173383	164319.0	0.0007	0.0071
14	122	109	1	108	200,000	198676	0.202376	158468.7	0.0007	0.0077
15	105	100	—	100	200,000	198568	0.231815	152537.0	0.0007	0.0084
16	104	96	1	95	200,000	198468	0.261386	146591.2	0.0006	0.0090
17	99	92	2	90	200,000	198373	0.290830	140680.1	0.0006	0.0097

18	124	115	2	113	200,000	198283	0.319935	134845.4	0.0008	0.0105
19	101	93	1	92	200,000	198170	0.348529	129102.0	0.0007	0.0112
20	83	76	1	75	200,000	198078	0.376479	123505.8	0.0006	0.0118
21	89	81	2	79	200,000	198003	0.403683	118072.5	0.0007	0.0125
22	81	75	1	74	200,000	197924	0.430065	112803.8	0.0007	0.0132
23	56	53	—	53	200,000	197850	0.455571	107715.3	0.0005	0.0137
24	57	53	2	51	200,000	197797	0.480165	102821.8	0.0005	0.0141
25	66	64	1	63	200,000	197746	0.503828	98116.0	0.0006	0.0148
26	39	38	—	38	180,000	177683	0.526551	84123.9	0.0005	0.0152
27	23	23	—	23	160,000	157645	0.548336	71202.6	0.0003	0.0156
28	24	22	—	22	140,000	137622	0.569192	59288.7	0.0004	0.0159
Total	2588	2414	14	2400	—	—	—	—	—	—

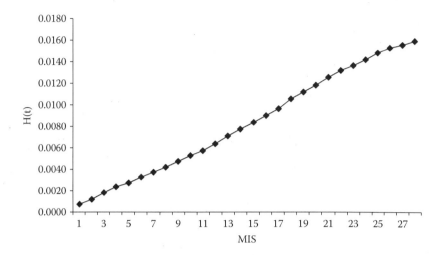

FIGURE 4.6   Hazard plot when mileage accumulation in vehicle population is known.

5 Hard Failures with Unknown Mileage Accumulation Rates*

Anybody can make history. Only a great man can write it.

—Oscar Wilde

OBJECTIVES

This chapter explains the following:

- Hazard function with a modification in the numerator, as the mileage accumulation rates in the vehicle population are not known
- Method for estimating the first failures outside the warranty mileage limit
- A five-step methodology for hazard rate estimation with an illustrated example

OVERVIEW

Section 5.1 gives the hazard rate for hard failures when mileage accumulation rates in the vehicle population are not known. Section 5.2 provides a method of estimating the first failures at each MIS (months in service) and a study of factors affecting the estimation process. Section 5.3 describes the five-step methodology, and Section 5.4 illustrates the methodology with an example.

5.1 HAZARD FUNCTION WITH MODIFICATION IN NUMERATOR

Sometimes the mileage accumulation rate for the vehicle population under study may not be available to the engineers carrying out the study. In such a situation, adjustments can be made to the numerator of Equation 3.1 (Chapter 3) to estimate the hazard rate at different MIS values. This can be done by estimating total number of first failures, say $n(t)$, expected at MIS $= t$. This includes the first claims observed within warranty coverage and estimated first failures beyond the warranty mileage limit. Equation 3.1 can now be written as follows:

---

* © 2003 Elsevier. Reprinted with permission from Elsevier: Rai, B. K. and Singh, N. (2003). Hazard rate estimation from incomplete and unclean warranty data. Reliability Engineering and System Safety, 81, 79–92.

$$h_3(t) = \frac{n(t)}{N(t)} \tag{5.1}$$

5.2 MODELING MILEAGE ON FAILED VEHICLES USING TRUNCATED NORMAL DISTRIBUTION

Let us consider a component-level failure for which the lognormal provides a good approximation to the distribution of mileage on the failed vehicles at a given MIS. Let Y be a random variable denoting the number of miles traveled by the failed vehicle. Then, $X = \ln(Y)$ follows the normal distribution. The probability density function (pdf) and cumulative density function (cdf) for normally distributed random variable X is, respectively, given by

$$f(x;\mu,\sigma) = \frac{1}{\sigma\sqrt{2\pi}} \exp\left[-\frac{1}{2}\left(\frac{x-\mu}{\sigma}\right)^2\right] \quad -\infty < x < \infty \tag{5.2}$$

and

$$F(x) = \int_{-\infty}^{x} f(w;\mu,\sigma)dw \tag{5.3}$$

Using the transformation $Z = \dfrac{X-\mu}{\sigma}$, the pdf and cdf of the standard normal distribution, respectively, become

$$\phi(z) = \frac{1}{\sqrt{2\pi}} \exp\left[-\frac{z^2}{2}\right] \quad (-\infty < z < \infty) \tag{5.4}$$

and

$$\Phi(z) = \int_{-\infty}^{z} \phi(t)dt \tag{5.5}$$

Let $\ln(M^l) = X_o$, which is the warranty mileage limit in terms of the normal variable. If an observation x_i is not included in the data set unless it is less than a specific value $X_o(x_i \le X_o)$, it is said to be singly truncated on the right at a point X_o. When the mean

$$\left(\overline{x_t} = \sum_{i=1}^{n} x_i / n \right)$$

is calculated from the truncated data, it is an underestimate of the population mean μ. Similarly, when the variance

$$\left(S_t^2 = \sum_{i=1}^{n} (x_i - \overline{x})^2 / [n-1] \right)$$

is obtained from the data, it is an underestimate of the population variance σ^2.

Consider a truncated distribution when the normal distribution with pdf (Equation 5.2) is truncated on the right at $x = X_o$. The pdf for the resulting truncated distribution is given by

$$f_{X_o}(x;\mu,\sigma) = \frac{\dfrac{1}{\sigma\sqrt{2\pi}} e^{\left[-\frac{1}{2}\left(\frac{x-\mu}{\sigma}\right)^2\right]}}{\dfrac{1}{\sigma\sqrt{2\pi}} \displaystyle\int_{-\infty}^{X_o} e^{\left[-\frac{1}{2}\left(\frac{x-\mu}{\sigma}\right)^2\right]} dx} \tag{5.6}$$

$$= \frac{1}{\sigma\sqrt{2\pi}\,F(X_o)} \exp\left[-\frac{1}{2}\left(\frac{x-\mu}{\sigma}\right)^2\right] \quad (-\infty < x \le X_o)$$

$$= 0 \text{ elsewhere}$$

The standardized pdf for the preceding expression is given by

$$\phi_{X_o}(z) = \frac{\phi(z)}{\Phi(\xi)} \quad (-\infty < z \le \xi) \tag{5.7}$$

$$= 0 \text{ elsewhere}$$

where $\xi = \dfrac{X_o - \mu}{\sigma}$ is the standardized point of truncation.

Hence, the problem of incomplete data set reduces to that of estimating the mean and standard deviation of a normally distributed population from a truncated data set when neither counts nor measurements of variates in the omitted portion are known. And then, these estimates can be used to obtain the total number of failures both within and outside the warranty mileage for vehicles at any

MIS. We can then use Equation 5.1 to arrive at an estimate of the hazard rate at each MIS.

5.2.1 METHODS TO ESTIMATE POPULATION PARAMETERS OF A TRUNCATED NORMAL DISTRIBUTION

To arrive at population parameters based on information from the truncated normal distribution, two most commonly used methods are

1. Two-moment estimator or maximum likelihood estimator (in the case of the normal distribution, both methods lead to identical estimators)
2. Three-moment estimators

Cohen (1959) developed two-moment estimators by equating the mean $\bar{x}$ and the variance s^2 of the truncated data set to the mean and variance of the truncated normal population. He considered the truncated distribution that results when the normal distribution with pdf (Equation 5.2) is truncated on the left at $x = X_o$. The resulting estimating equations were solved to arrive at estimates for population mean μ and the population variance σ^2 using equations

$$\hat{\sigma}^2 = s^2 + \hat{\theta}(\bar{x} - X_o)^2 \text{ and } \hat{\mu} = \bar{x} - \hat{\theta}(\bar{x} - X_o) \tag{5.8}$$

Cohen and Whitten (1988) developed tables to aid such an estimation with minimum need for calculations by providing values of $\hat{\theta}$ for different values of α, where

$$\hat{\alpha} = \frac{s^2}{(\bar{x} - X_o)^2}$$

The benefit of using three-moment estimators lies in estimating the parameters of the normal distribution without resort to special tables. The k-th sample moment of a singly left-truncated sample at X_o about the left terminus is given as

$$v_k' = \sum_{i=1}^{n} \frac{(x_i - X_o)^k}{n} \tag{5.9}$$

The estimates of population mean μ and the population variance σ^2 were derived by Cohen (1951) and are given as

$$(\sigma^2)^* = \frac{(v_2')^2 - v_1' v_3'}{v_2' - 2(v_1')^2} \tag{5.10}$$

and

$$\mu^* = X_o + a^* \tag{5.11}$$

where $a^* = \dfrac{v_3' - 2v_1'v_2'}{v_2' - 2(v_1')^2}$.

The previous equations are derived assuming that the truncation occurred in the left tail of the distribution. They are likewise applicable when truncation is in the right tail, in which case the odd moments are negative owing to the choice of origin.

5.2.2 A STUDY OF FACTORS AFFECTING THE ESTIMATION PROCESS

In practical applications, deviations from normality and varying amounts of truncation of the original distribution are expected to have an impact on the efficiency of estimation. In this section, a simulation-based experiment studies the behavior of estimation methods described in Section 5.2.1. For this purpose, various sets of observations are randomly generated, and each data set is treated as a complete data set. From these complete data sets, truncated data sets are artificially created for the study.

Experimental factors and levels are given in Table 5.1. Factor A has already been discussed in Section 5.2.1. Factor B is used here to capture the effect of departure of normality on the estimation process. For this, significance level "p" based on the Anderson–Darling (AD) statistic is used (Kececioglu 1993). Note that even when the data sets are randomly generated from the same distribution, p-value based on AD statistic can lie between 0 and 1 because of sampling fluctuations. A value close to 1 indicates closeness of the complete data set to the chosen distribution, and vice versa. Factor C has three levels; singly left truncated at $-1.5s$, $-1.0s$, and $-0.5s$, reflecting approximately 7%, 16%, and 31% truncation of the complete data set, respectively (where s is the standard deviation based on complete data set). Thus, the total possible combinations for the factors and levels are $2 \times 3 \times 3 = 18$. Because this is a simulation-based experiment, to run all 18 combinations a full factorial experimental design is chosen.

In all, 12 distinct "complete" data sets were constructed. This comprised four replicates at each of the three levels of Factor B. Each data set consists of 3000 observations randomly generated from SND(0,1).

To compute the efficiency of predicting count of units in a complete data set, the following procedure was used. Let n be the count of units in the truncated data set, and assume N, denoting corresponding number for the complete data set, is unknown and needs to be estimated. Using the two methods described, population parameters μ and σ^2 can be easily estimated, which then can be used for providing an estimate $\hat{N}$ for N using the formula given as follows:

$$\hat{N} = \frac{n}{P(X \geq X_o)} \tag{5.12}$$

TABLE 5.1

Factors and Levels for a Simulation-based Experiment for Studying the Factors Influencing the Estimation Process

Factor	Level 1	Level 2	Level 3
A. Estimation method	Two-moment estimator	Three-moment estimator	—
B. p based on AD statistic	$p > 0.9$	$p \sim 0.5$	$p < 0.1$
C. Truncated on left at	$-1.5s$	$-1.0s$	$-0.5s$

Now using $N = 3000$ for this experiment and $\hat{N}$ estimated based on an artificially truncated data set, a measure for efficiency of estimation method can be given as

$$\text{Efficiency}(\%) = 100 - \left(\frac{|N - \hat{N}|}{N} \right) \times 100 \qquad (5.13)$$

Efficiency would be 100% when $N = \hat{N}$, that is, when the method used leads to estimated value to be the same as the actual value. As the error between actual N and the estimate $\hat{N}$ increases, the efficiency decreases.

Figure 5.1 gives a summarized parameter diagram for the simulation-based experiment. The process of estimating parameters of a truncated normal distribution is at the center of the p-diagram. Input is in the form of randomly generated observations from the standard normal distribution. Control factors are those that can be changed by the experimenter and are as given in Table 5.1. Noise factors are those over which the experimenter does not have much control. The ideal function of the process under study represents the situation where the efficiency of predicting count of units in the complete data set is perfect. However, the error state of the response

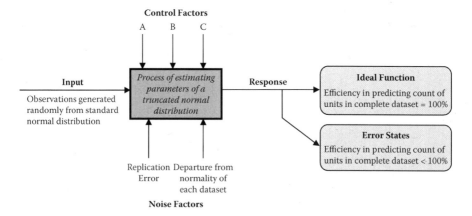

FIGURE 5.1 Parameter diagram of the simulation-based experiment.

cannot be totally avoided in practical situations owing to the presence of noise factors, and thus, the aim is to minimize the impact of noise factors.

For the calculations involving the two-moment estimator, Equation 5.8 is used to arrive at estimates for the population mean and variance. Subsequently, efficiency is arrived at using Equations 5.12 and 5.13. For the three-moment estimator method, Equations 5.10 and 5.11 are used to estimate the population mean and variance, after which, the value of efficiency is calculated in a similar manner to the first method. The full factorial experimental layout and results obtained are given in Table 5.2.

Analysis of variance was carried out to identify significant factors affecting the efficiency of the estimation process under study. Table 5.3 gives the analysis of variance (ANOVA) table (Montgomery 1991). Statistically significant factors at the 1% level of significance or below can be quickly identified from the table.

It can be seen from Table 5.3 that the main effects B and C and interaction B × C are found to be statistically significant. This indicates that departure from normality of the complete data set and amount of truncation, individually as well as jointly, have significant influence on the efficiency with which the count of units in the complete data set can be estimated. Factor A is not significant, indicating that any of the two methods can be used in estimating N without in any way compromising on the efficiency of estimation. Average response graphs for significant factors and interaction are shown in Figure 5.2.

From Figure 5.2, we observe that for the main effects B and C, the efficiency of estimating N for the complete data set is seen to come down when both depart from normality and amount of truncation increase. The response graph for interaction B × C sheds more light on the behavior. From the interaction graph for B × C, we observe that as we move from B1 to B3, the downward slopes for C1, C2, and C3 become steeper. The drop in efficiency, especially for C3, is quite steep between B2 and B3. In other words, when the departure from normality is low to moderate, efficiency of estimation process does not fall as sharply for increasing amount of truncation as it does when there is high departure from normality. This calls for a word of caution in situations where significant departure from normality may be expected.

Two important observations can be made from this experiment:

1. There is no major difference between the two methods of estimation: two-moment estimator and three-moment estimator. Thus, we can make use of the three-moment estimator when the special tables required for calculations are not readily available, or even otherwise.
2. The results showed that the estimation methods under study are significantly influenced by amount of truncation. However, sufficiently high efficiency can still be achieved in estimating count of units in the complete data set, when departure from normality is not statistically significant at the 10% level of significance or below based on AD statistic.

TABLE 5.2
Full Factorial Experimental Layout and Results

Number	Estimation Method (A)	p Based on AD Statistic (B)	Truncated at (C)	Response Efficiency of Eestimating N			
1	2-Moment	>0.9	−1.5s	100.0	99.3	100.0	99.8
2	2-Moment	>0.9	−1.0s	98.7	99.7	99.6	99.5
3	2-Moment	>0.9	−0.5s	97.1	98.9	99.5	99.3
4	2-Moment	~0.5	−1.5s	99.7	99.6	99.5	98.5
5	2-Moment	~0.5	−1.0s	97.3	99.4	99.3	98.2
6	2-Moment	~0.5	−0.5s	97.3	98.6	98.1	99.7
7	2-Moment	<0.1	−1.5s	98.4	98.8	99.7	99.9
8	2-Moment	<0.1	−1.0s	95.2	98.3	97.0	99.6
9	2-Moment	<0.1	−0.5s	93.9	93.1	96.4	99.3
10	3-Moment	>0.9	−1.5s	99.8	99.3	99.9	99.8
11	3-Moment	>0.9	−1.0s	98.4	99.4	99.8	99.1
12	3-Moment	>0.9	−0.5s	96.4	98.2	99.2	98.4
13	3-Moment	~0.5	−1.5s	99.8	99.8	100.0	98.4
14	3-Moment	~0.5	−1.0s	98.7	99.4	98.3	97.9
15	3-Moment	~0.5	−0.5s	95.8	99.7	95.9	98.7
16	3-Moment	<0.1	−1.5s	97.7	98.4	99.8	99.9
17	3-Moment	<0.1	−1.0s	94.8	97.0	98.5	98.9
18	3-Moment	<0.1	−0.5s	93.8	90.4	95.0	95.6

TABLE 5.3

Analysis of Variance for the Simulation Study

Source	SS	DF	MS	F	p
A	3.705	1	3.705	2.0	0.158
B	56.239	2	28.120	15.5	0.000
C	69.267	2	34.634	19.1	0.000
AB	1.307	2	0.654	0.4	0.698
AC	4.669	2	2.335	1.3	0.283
BC	25.541	4	6.385	3.5	0.012
ABC	1.085	4	0.271	0.2	0.962
Error	97.663	54	1.809	—	—
Total	259.477	71	—	—	—

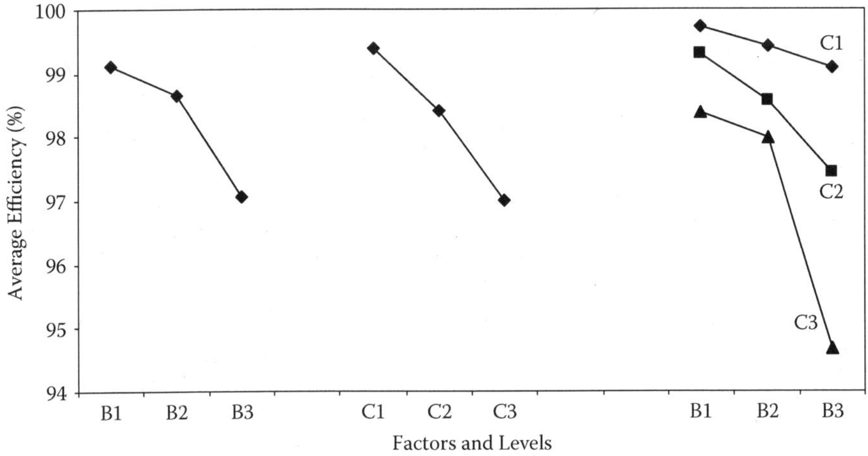

FIGURE 5.2 Average response graph for efficiency in estimating N.

5.3 A FIVE-STEP METHODOLOGY

A five-step methodology to arrive at the hazard rate estimate given by Equation 5.1 is shown in Figure 5.3.

Step 1: For every vehicle, warranty claims with minimum mileage are retained and repeat claims are screened out. The purpose is to keep manufacturing quality separate from service or repair quality. Misclassified and misdiagnosed claims are identified through one or more of the methods discussed in Chapter 3 and excluded from further analysis (same as Step 1 in Chapter 4, Section 4.3).

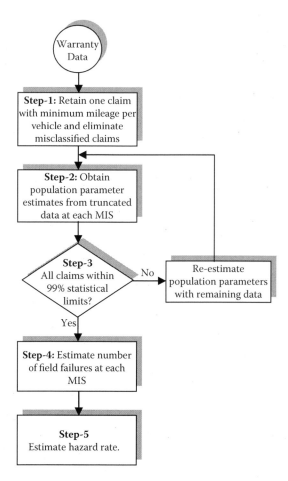

FIGURE 5.3   Flowchart of the five-step methodology for estimation of hazard function from hard failures with unknown mileage accumulation rates in vehicle population.

Step 2: This step deals with the issue of incompleteness of warranty data. To obtain hazard rate estimates at each MIS, an estimate of total vehicles failed, both within and outside the warranty mileage, is required. This requires parameter estimation of the complete data set. The first claims at each MIS are treated as right-truncated data set and population parameters of the complete distribution are estimated using Equations 5.10 and 5.11. Note that in the case of the lognormal distribution, natural logarithmic transformation of miles traveled at failure is carried out before using moment estimators. When using the normal distribution, no transformation is required.

In this step, distributional assumption of data following lognormal or normal distribution is also verified. Graphical goodness-of-fit test using probability plots is used for this purpose. If a *p*-value based on AD statistic is found to be >0.1, it can be concluded that the departure from the assumed distribution is not statistically significant. We propose to leave out vehicles in their first MIS for verifying the distributional assumption, as they are likely to be influenced by the presence of vehicles escaping the final test for a given failure mode. Data from 2 to 5 MIS can be used for the purpose, as for these months the data are unlikely to be affected by the incompleteness problem, as the probability of a vehicle accumulating more than 36,000 mi is extremely remote, if not zero. If the distributional assumption is not found valid from the probability plots, it is advisable to look for other appropriate methods.

Step 3: Varying usage rates can sometimes give rise to unusual mileage accumulation by a customer. Such unusual observations can lead to inflated figures for population parameter estimates. Hence, in this step unusual observations that are outside the 99% statistical limits are identified and population parameters are reestimated with the remaining data. This ensures that stabilized estimates for population parameters are obtained. The data points found to be outside the 99% statistical limits are still included in the total number of first failures when estimating the hazard rate in Step 5.

Step 4: Using the final population parameter estimates and the number of observations in the truncated warranty data set n, the estimated number of vehicles failing in the field for the first time is obtained using Equation 5.12.

Step 5: The hazard rate is estimated using Equation 5.1.

5.4   AN APPLICATION EXAMPLE

This application example uses the same data used in Chapter 4. It is assumed that the mileage accumulation rate in the vehicle population is not known.

Step 1: For a component-level failure mode of a vehicle, 2574 claims were reported during MIS = 1 to 28. After completing the first step, 160 repeat claims were discarded. Also, using the information in the technician comments and returned parts analysis, 14 claims were found to be misclassified and were therefore excluded. The remaining 2400 claims were considered for further analysis. (This step is the same as Step 1 in Chapter 4, Section 4.4.)

Step 2: Probability plots of mileage to first warranty claim for the first 5 months from the date of sale are given in Figure 5.4. We observe as expected that but for the first MIS, the points lie very close to a straight line for the remaining 4 months, indicating a satisfactory fit to the lognormal distribution. It is also verified that the p-value based on AD statistic, except for the first month, is >0.10, confirming that the departure from the lognormal distribution is not statistically significant.

For the estimation of population parameters, consider vehicles with MIS = 15, when the probability of a vehicle failing outside the warranty mileage limit of 36K is sufficiently high. Let Y denote the miles traveled by the vehicle. Assuming the lognormal distribution for Y and using the transformation

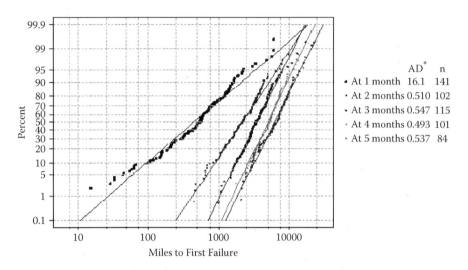

FIGURE 5.4   Lognormal probability plot of miles to first failure for first 5 months.

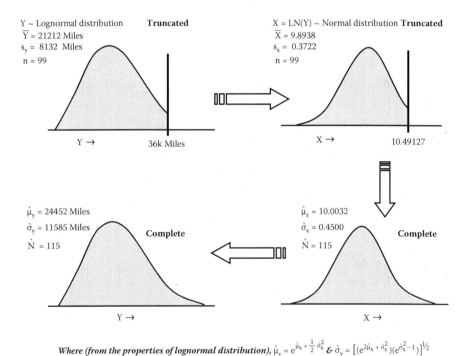

FIGURE 5.5   The process of estimating field failures at MIS = 15.

$X = \ln(Y)$, the pdf and standardized pdf of the truncated normal distribution will be as given in Equations 5.6 and 5.7, respectively, where $X_o = \ln(36000)$ = 10.49127. Figure 5.5 depicts the process of estimating total field failures for MIS = 15.

Step 3: A total of 55 claims during MIS = 1 to 28, ranging from 0 to 5 at any MIS, were found outside 99% statistical limits. For example, at MIS = 15, there were 100 vehicles with first warranty claim against the specific failure mode. However, one vehicle with a mileage of about 3919 mi is found outside the 99% statistical limits. A mileage accumulation of just 260 mi/month is highly unusual when the remaining vehicles in the group had an average above 1400 mi/month. To avoid highly unstable estimates of population means and variance, the final population estimates were based on $n = 99$. The process predicted 115 field failures for vehicles at MIS = 15, as shown in Figure 5.5. The effectiveness of this step in arriving at more stable estimates for the population parameter can be judged by comparing lognormal probability plots based on raw warranty data after excluding data points outside the 99% statistical limits. Figure 5.6 shows one such comparison for failures at MIS = 6.

Step 4: Out of a total of 2400 first claims, 51 were found to be outside the 99% statistical limits at various MIS values. Using 2349 claims within the warranty coverage, all 2831 first field failures were estimated, excluding the claims outside the 99% limit, as shown in Table 5.4.

It can be seen from Table 5.4 that $n(t)$ includes the claims found outside 99% statistical limits. $N(t)$ is obtained by subtracting the total number of first failures in $n(t)$ column up to $t - 1$ from V_t.

Step 5: From the hazard rate estimates obtained using Equations 5.1, a hazard plot (shown in Figure 5.7) was prepared.

We observe from Figure 5.7 that the cumulative hazard rate largely shows a constant failure rate (CFR) pattern.

5.5 SUMMARY

Key points from the chapter are summarized in the following:

- When the mileage accumulation rate in the vehicle population is not known, the mileage distribution of the first claims at each MIS value is treated as right truncated at the mileage limit M_1. As the process of estimating the population parameters uses each reported mileage value, it is highly sensitive to the presence of unusually low mileage values. Thus, estimating the population parameter of a complete distribution based on mileage values within the 99% statistical limits helps to provide stable estimates for the number of first failures outside the warranty mileage limit.
- Using a simulation-based experiment, the efficiency of two-moment and three-moment estimators to estimate the population parameters of a truncated data set was compared. It showed comparable efficiencies for the two

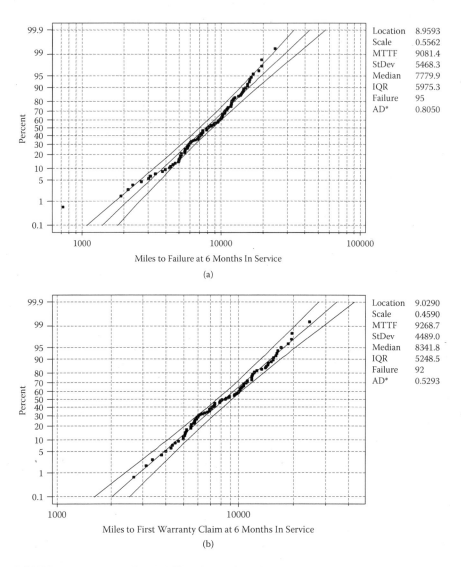

FIGURE 5.6 Lognormal probability plots (a) based on raw warranty data, (b) after excluding data points outside the 99% statistical limits.

estimators. Thus, one can make use of the three-moment estimator when the special tables required for calculations are not readily available, or even otherwise. The results further showed that estimation methods under study are significantly influenced by the amount of truncation in the data set. However, sufficiently high efficiency can still be achieved in estimating count of units in the complete data set, when departure from normality is not statistically significant.

- Unavailability of mileage accumulation rates in the vehicle population greatly increases the computational efforts needed to arrive at hazard rate

TABLE 5.4

Hazard Rate Estimates for Hard Failures with Unknown Mileage Accumulation Rates in Vehicle Population

MIS (t)	Number of Valid First Claims	Claims Outside 99% Limit	Claims Within 99% Limit	Estimated First Failures, Excluding Claims Outside 99% Limit	Estimated First Failures $n(t)$	Number of Vehicles in Field V_t	$N(t)$	$h(t)$	$H(t)$
1	141	0	141	141	141	200,000	200,000	0.00071	0.00071
2	103	0	103	103	103	200,000	199,859	0.00052	0.00122
3	117	0	117	117	117	200,000	199,756	0.00059	0.00181
4	102	0	102	102	102	200,000	199,639	0.00051	0.00232
5	84	0	84	84	84	200,000	199,537	0.00042	0.00274
6	95	3	92	92	95	200,000	199,453	0.00048	0.00321
7	101	2	99	99	101	200,000	199,358	0.00051	0.00372
8	79	3	76	76	79	200,000	199,257	0.00040	0.00412
9	99	1	98	98	99	200,000	199,177	0.00050	0.00462
10	95	1	94	96	97	200,000	199,078	0.00049	0.00510
11	88	1	87	90	91	200,000	198,982	0.00046	0.00556
12	111	0	111	114	114	200,000	198,890	0.00057	0.00613
13	109	2	107	115	117	200,000	198,777	0.00059	0.00672
14	108	5	103	112	117	200,000	198,660	0.00059	0.00731
15	100	1	99	115	116	200,000	198,543	0.00058	0.00789

continued

TABLE 5.4 (continued)

Hazard Rate Estimates for Hard Failures with Unknown Mileage Accumulation Rates in Vehicle Population

MIS (t)	Number of Valid First Claims	Claims Outside 99% Limit	Claims Within 99% Limit	Estimated First Failures, Excluding Claims Outside 99% Limit	Estimated First Failures $n(t)$	Number of Vehicles in Field V_t	$N(t)$	$h(t)$	$H(t)$
16	95	1	94	143	144	200,000	198,427	0.00073	0.00862
17	90	4	86	96	100	200,000	198,283	0.00050	0.00912
18	113	1	112	152	153	200,000	198,183	0.00077	0.00990
19	92	2	90	129	131	200,000	198,030	0.00066	0.01056
20	75	3	72	84	87	200,000	197,899	0.00044	0.01100
21	79	4	75	152	156	200,000	197,812	0.00079	0.01179
22	74	2	72	112	114	200,000	197,656	0.00058	0.01236
23	53	2	51	98	100	200,000	197,542	0.00051	0.01287
24	51	4	47	58	62	200,000	197,442	0.00032	0.01318
25	63	3	60	134	137	200,000	197,380	0.00070	0.01388
26	38	3	35	41	44	180,000	177,242	0.00025	0.01413
27	23	1	22	48	49	160,000	157,198	0.00031	0.01444
28	22	2	20	28	30	140,000	137,149	0.00022	0.01466
Total	2400	51	2349	2831	2882	—	—	—	—

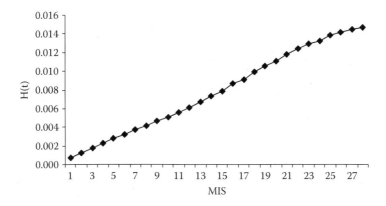

FIGURE 5.7 Hazard plot when mileage accumulation in vehicle population is not known.

estimates. Kalbfleisch and Lawless (1988) note that in the absence of mileage accumulation rates, for the truncated warranty data to be informative, a sufficiently high proportion of items should fail before M_1. However, in contrast to a warranty mileage limit of about 12K mi a decade ago, when the article was published, today mileage limits of 36K or more are not uncommon.

BIBLIOGRAPHIC NOTES

Estimation of population parameters for truncated distributions in general and truncated normal distributions in particular has been widely studied. Galton (1897) studied the truncated normal distribution in relation to speeds of horses and estimated mean from mode obtained by plotting frequency polygons. He also estimated population standard deviation using interquartile ranges by equating mode to median. Pearson (1902) proposed another procedure for estimating normal distribution parameters from truncated samples, by fitting parabolas to logarithms of the sample frequencies. Later Pearson and Lee (1908) used the method of moments to estimate the mean and standard deviation of a normal distribution from a singly left-truncated sample. Fisher (1931) employed the method of maximum likelihood to derive estimators based on singly truncated samples for normal distribution parameters. Fisher also derived asymptotic variances and covariances for his estimators. Stevens (1937), Cohen (1949), Gupta (1952), Raj (1952), Swamy (1962), Zakkula (1966), Steve (1976), and others extended the work further, including estimation based on restricted data from other distributions. Cohen (1991) consolidated the work by incorporating scattered results into a single volume.

6 Soft Failures with Known Mileage Accumulation Rates*

The best things in life are censored.

—**Woody Allen**

OBJECTIVES

This chapter explains the following:

- The phenomena of customer-rush near warranty expiration limit
- Hazard rate estimate formulas for months in service (MIS) and mileage as life variables for soft failures with a possibility of delayed failure reporting
- Method for determining artificial truncation points a and $M_1 - b$
- A method to incorporate left and right censoring in warranty data when estimating hazard rates
- Confidence interval for the hazard rate estimates

OVERVIEW

This chapter provides a methodology to arrive at hazard rate estimates for failure modes belonging to the soft failure category. It specifically focuses on soft failure modes that have a high chance of detection by users and whose reporting is delayed until the vehicle's warranty coverage is about to expire. The chapter covers the case when the mileage accumulation in the vehicle population is known or can be estimated. Section 6.1 describes the occurrence of spikes at the beginning and toward the end of the warranty period. Sections 6.2 and 6.3 give formulas to estimate hazard rates for MIS and mileage as life variables, respectively. Section 6.4 provides a method to incorporate censoring information during hazard rate estimation. Section 6.5 gives a flowchart of the overall methodology. Section 6.6 illustrates the methodology with an example and comments on the results.

---

* © 2006 IEEE. The major portion of this chapter is reprinted with permission from Rai, B.K. and Singh, N. (2006). Customer-rush near warranty expiration limit and non parametric hazard rate estimation from known mileage acculaltion rates. IEEE Transactions on Reliability, 55(3), 480–489.

6.1   INTRODUCTION

Many automobile users delay the reporting of soft failures (failures that result in degraded performance but leave the vehicle operable) until the very end of the warranty period. Such a delay in reporting often leads to the occurrence of "spikes" in warranty claims toward the end of the warranty period. On the other hand, occurrence of quality issues (manufacturing or assembly defects) in addition to reliability issues (usage-related failures) often lead to spikes in warranty claims at the beginning of the warranty period (Majeske 2003). A few examples of such a phenomenon are depicted in Figure 6.1, which shows histograms for the number of vehicles with first warranty claim in mileage bands of 1K mi ("K" represents mileage in thousands). The first bar with 0.5K mi as midpoint represents the number of first warranty claims with recorded mileage >0 and ≤1K mi. The second bar represents number of first claims with recorded mileages >1K and ≤2K mi, and so on. The six plots in Figure 6.1 represent three different failure modes coded as "1," "2," and "3"

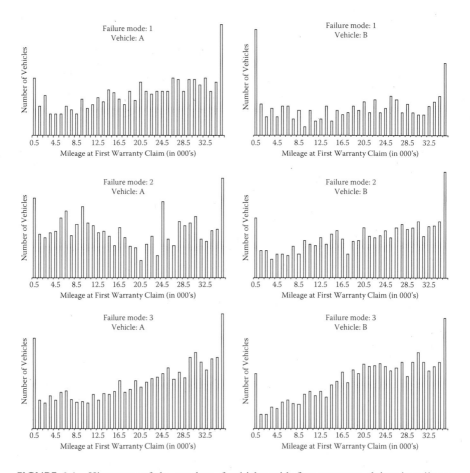

FIGURE 6.1   Histogram of the number of vehicles with first warranty claims in mileage bands of 1K mi.

and two different vehicles coded as "A" and "B." A common warranty period of a minimum of 3 years or 36K mi, whichever occurs first, applies to all six plots.

A common feature in all six plots is the existence of spikes in warranty claims at the beginning and near the end of the warranty coverage period. The spike in warranty claims observed in the mileage band of 35K–36K mi is about twice the observed number of first claims in the neighboring mileage bands, which makes it look very unnatural. It is difficult to believe that such a spike within 1K mi could systematically occur for different failure modes and different vehicle lines, owing to natural increase in usage-related vehicle failures. It definitely points toward the existence of some special cause. One plausible explanation is the existence of customer-rush near the warranty expiration limit resulting in spikes observed toward the end of the warranty mileage limit. On the other hand, the spike in claims for vehicles with <1K mi indicates the existence of quality issues in addition to the reliability issues at the initial usage period of a vehicle (Rai and Singh 2006).

When carrying out reliability analysis using warranty data, it is important to address such events to provide accurate representation of the field failures to design engineers. Proper feedback on field failures would enable design engineers to plan and take appropriate design improvement actions for current as well as forward model-year vehicles.

6.2 RISK SET ADJUSTMENT: MIS AS LIFE VARIABLE

The denominator in Equation 4.1 (Chapter 4) represents the number of vehicles at risk of first claim within the warranty mileage limit of M_1 miles at MIS = t. To avoid mixing of manufacturing- or assembly-related claims with usage-related claims, the claims below, say, a miles, at each MIS may be screened out. Similarly, claims occurring in the last b miles of the warranty mileage limit M_1 miles may be screened out to avoid bias in the estimation of the hazard function. Such artificial truncation of the warranty data set would require further modification of Equation 4.1 when estimating the hazard rate. Let W_t (t = 1, 2, 3, ..., M_2) be a random variable denoting the miles driven by a vehicle at MIS = t, and let n_t denote the number of first claims between a miles and $M_1 - b$ miles at MIS = t. We can now rewrite Equation 4.1 as follows:

$$h_4(t) = \frac{n_t}{N(t) \times P[a \leq W_t \leq (M_1 - b)]} \tag{6.1}$$

This equation treats the warranty data at each MIS to be left truncated at a miles and right truncated at $M_1 - b$ miles. The denominator of Equation 6.1 represents the number of vehicles at risk of first claim in the interval (a, $M_1 - b$) miles at each MIS. The artificial truncation points a and $M_1 - b$ are depicted in Figure 6.2.

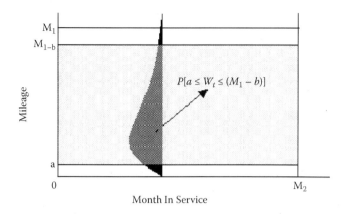

FIGURE 6.2 Risk set adjustment when mileage distribution in the vehicle population is known.

EXAMPLE 6.1

Consider Example 4.1 (Chapter 4), in which $W \sim LN(7.279, 0.688^2)$. Also, at 15 MIS, $n_{15} = 41$, $N(15) = 120{,}000$, $M_1 = 36K$ mi, and $a = b = 1K$. Then, using $W_{15} \sim LN[7.279 + \ln(15), 0.688^2]$, we get $P[1K < W_{15} < 35K] = 0.755$. Thus, the number of vehicles at risk of first claim in the interval $(a = 1K, M_1 - b = 35K)$ is $0.755 \times 120{,}000 = 90{,}600$. Accordingly, using Equation 6.1, we get $h_4(15) = 0.000453$.

6.3 RISK SET ADJUSTMENT: MILEAGE AS LIFE VARIABLE

Let m_i denote the number of first claims in the mileage interval $(k_{i-1}, k_i]$ with $k_1 = 0$ and $i = 2, 3, \ldots$, and so on. Mileage bands can be of 1000 mi such as $(0, 1000]$, $(1000, 2000]$, and so on. $V_t - V_{t+1}$ gives the number of vehicles that are exactly t MIS at the time of data analysis. The number of vehicles that are right censored up to $(i - 1)$ mileage interval is given by

$$P\left[W_t \leq k_{i-1}\right]\left(V_t - V_{t+1}\right) \tag{6.2}$$

The total number of vehicles at MIS, $t = 1, 2, \ldots, M_2$ that are right-censored up to mileage interval $(i - 1)$ is given by

$$\sum_{t=1}^{M_2} P\left[W_t \leq k_{i-1}\right]\left(V_t - V_{t+1}\right) \tag{6.3}$$

Thus, the total number of vehicles at MIS $= 1, 2, \ldots, M_2$ that are right censored at the end of mileage interval $(i - 1)$ represented by r_i is given as

$$r_i = \sum_{t=1}^{M_2} P\left[W_t \leq k_{i-1}\right]\left(V_t - V_{t+1}\right) - \sum_{t=1}^{M_2} P\left[W_t \leq k_{i-2}\right]\left(V_t - V_{t+1}\right) \tag{6.4}$$

An estimate of the hazard function for the i-th mileage interval can thus be arrived at by using

$$h_5(i) = \frac{m_i}{N_i - r_i} \tag{6.5}$$

where $N_i = N_{i-1} - m_{i-1}$ for $i = 2, 3, \ldots$; $N_1 = N$. N_i is the number of vehicles without claims at the beginning of the mileage interval i.

EXAMPLE 6.2

Consider mileage bands of 1000 mi, and let values of V_t be as given in Table 6.1. If $W_t \sim LN[7.279 + \ln(t), 0.688^2]$, the calculations to obtain r_{10} are given here.

Table 6.1 shows that overall there are 200,000 vehicles in the field. Of them, 10,000 vehicles have reached 36 MIS, whereas all 200,000 vehicles have reached 25 MIS.

From Table 6.1, we obtain

$V_t - V_{t+1} = 0$ for $t = 1, 2, \ldots, 24$
$V_t - V_{t+1} = 20,000$ for $t = 25, 26, \ldots, 31$
$V_t - V_{t+1} = 15,000$ for $t = 32$ and 33
$V_t - V_{t+1} = 10,000$ for $t = 34, 35,$ and 36

$$r_{10} = \sum_{t=1}^{M_2} P\left[W_t \leq k_9\right]\left(V_t - V_{t+1}\right) - \sum_{t=1}^{M_2} P\left[W_t \leq k_8\right]\left(V_t - V_{t+1}\right)$$

$= 2518.5 - 1605.9$
$= 912.6$
~ 913 vehicles

TABLE 6.1

Number of Vehicles in the Field at Each MIS Value

MIS (t)	Vt	MIS (t)	Vt	MIS (t)	Vt	MIS (t)	Vt
1	200,000	11	200,000	21	200,000	31	80,000
2	200,000	12	200,000	22	200,000	32	60,000
3	200,000	13	200,000	23	200,000	33	45,000
4	200,000	14	200,000	24	200,000	34	30,000
5	200,000	15	200,000	25	200,000	35	20,000
6	200,000	16	200,000	26	180,000	36	10,000
7	200,000	17	200,000	27	160,000		
8	200,000	18	200,000	28	140,000		
9	200,000	19	200,000	29	120,000		
10	200,000	20	200,000	30	100,000		

Thus, it is estimated that approximately 913 vehicles that entered the mileage interval (8000, 9000] would not exceed a mileage of 9000 mi. Thus, 913 vehicles are estimated to be right censored at the end of the 9th mileage interval or at the beginning of the 10th mileage interval.

6.4 INCORPORATING CENSORING INFORMATION IN THE HAZARD FUNCTION ESTIMATION

When claims data are doubly truncated and mileage accumulation rates are available, Equation 6.1 provides an estimate of the hazard function. However, on certain occasions, data on left-censored first claims due to delayed reporting of soft failures during $(M_1 - b, M_1)$ can be identified using methods discussed in Chapter 2.

A left-censored first warranty claim occurring at MIS = t requires an adjustment to the total number of first claims for each MIS $\leq t$. To enable such an adjustment for estimating the hazard function, we make use of an iterative procedure developed by Turnbull (1974). This procedure is based on the product-limit method and also uses the idea of self-consistency. The method also incorporates right-censored cases of vehicles that are in the field and have not yet failed. The procedure helps to obtain maximum likelihood estimates of the reliability function. The method is useful when the data can be grouped naturally, as in the case of automobile warranty data, where claims are grouped by MIS values. Although the main focus of the methodology is on MIS as a life variable, the transition to mileage as life variable is straightforward.

6.4.1 NONPARAMETRIC MLE OF HAZARD FUNCTION

As warranty data is generally both truncated and censored, the maximum likelihood estimator (MLE) is one of the most popular methods of estimating population parameters (Kalbfleisch and Lawless 1988, 1992; Cohen 1991). MLEs were initially formulated by C. F. Gauss but were first introduced as a general method of estimation by Prof. R. A. Fisher in a series of papers (Gupta and Kapoor 1989).

Maximum likelihood estimating equations are generally obtained by equating first partial derivatives of the loglikelihood function with respect to the parameters, to zero. The advantage of using MLEs includes ease of obtaining variance–covariance matrix of the estimates. The variance–covariance matrix is the inverse of the Fisher information matrix containing the negative of expected values of second partial derivatives of the loglikelihood function.

Let $R(t) = P[T > t]$ be the reliability function at $t = 1, 2, \ldots, M_2$. As defined earlier, n_t represents the number of first claims in the mileage interval $(a, M_1 - b)$ miles at MIS = t. Each of these n_t reported failures contributes a term $[R(t-1) - R(t)]$ to the likelihood function. Let r_t denote the number of vehicles that have not yet completed t MIS and are right censored at MIS = t. Each of these r_t right-censored vehicles contributes a term $R(t)$ to the likelihood function. Similarly, let c_t be the number of left-censored first claims at MIS = t that contribute a term $[1 - R(t)]$ to the likelihood function.

$$N = \sum_{j=1}^{M_2} \left(n_j + r_j + c_j \right)$$

is the total number of vehicles in the field. To obtain nonparametric maximum likelihood estimates of $R(t)$ (Meeker and Escobar 1998), the loglikelihood function L can be written as

$$L = \ln \left(\prod_{t=1}^{M_2} [R(t-1) - R(t)]^{n_t} [R(t)]^{r_t} [1 - R(t)]^{c_t} \right)$$

$$= \sum_{t=1}^{M_2} \left\{ n_t \ln[R(t-1) - R(t)] + r_t \ln[R(t)] + c_t \ln[1 - R(t)] \right\}$$

(6.6)

Using the results obtained by Turnbull (1974), we can obtain the maximum likelihood estimates $\hat{R}(t)$ for the reliability functions in Equation 5.6 (Chapter 5) as

$$R(t) = q_1 \times R(t-1), \; t = 2, 3, \ldots, M_2$$

(6.7)

With $R(0) = 1$, $R(1) = q_1$, and

$$q_t = \frac{\left(N'(t) - n_t' \right)}{N'(t)}$$

(6.8)

$$N'(t) = P\left[a \leq W_t \leq (M_1 - b)\right] \times \sum_{i=t}^{M_2} (r_i + n_i')$$

(6.9)

$$n_t' = n_t + \sum_{i=t}^{M_2} c_i \alpha_{ij}$$

(6.10)

$$\alpha_{it} = \frac{R(t-1) - R(t)}{1 - R(i)}, t \leq i$$

(6.11)

The iteration procedure starts with obtaining $R(i^0)$ as the initial estimate, assuming $c_i = 0, \forall i = 1, 2, \ldots, M_2$. Then, using Equations 6.8–6.11, values for $R(i^l), l \geq 0$, are iteratively obtained unti $\max_{1 < i < M_2} \left| R(i^l) - R(i^{l-1}) \right| < \delta$, where δ is a very small number. From the estimates $\hat{R}(t)$, estimates of the cumulative hazard function $\hat{H}(t)$ are obtained from the following relationship:

$$\hat{H}(t) = -\ln[\hat{R}(t)], t = 1, 2, 3, \dots M_2 \tag{6.12}$$

6.4.2 VARIANCE–COVARIANCE MATRIX FOR THE MAXIMUM LIKELIHOOD ESTIMATES

An approximate variance–covariance matrix of $\hat{R}(t)$, $t = 1, 2, 3, \dots, M_2$, is obtained as an inverse of the Fisher's information matrix, that is,

$$
\begin{bmatrix}
Var[\hat{R}(1)] & Cov[\hat{R}(1),\hat{R}(2)] & 0 & \cdots & 0 & 0 \\
Cov[\hat{R}(1),\hat{R}(2)] & Var[\hat{R}(2)] & Cov[\hat{R}(2),\hat{R}(3)] & \cdots & 0 & 0 \\
0 & Cov[\hat{R}(2),\hat{R}(3)] & Var[\hat{R}(2)] & \cdots & 0 & 0 \\
\cdot & \cdot & \cdot & & \cdot & \cdot \\
\cdot & \cdot & \cdot & & \cdot & \cdot \\
\cdot & \cdot & \cdot & & \cdot & \cdot \\
0 & 0 & 0 & \cdots & Var[\hat{R}(M_2-1)] & Cov[\hat{R}(M_2-1),\hat{R}(M_2)] \\
0 & 0 & 0 & \cdots & Cov[\hat{R}(M_2-1),\hat{R}(M_2)] & Var[\hat{R}(M_2)]
\end{bmatrix}
$$

$$
=
\begin{bmatrix}
-\dfrac{\partial^2 L}{\partial[R(1)]^2} & -\dfrac{\partial^2 L}{\partial R(1)\partial R(2)} & 0 & \cdots & 0 & 0 \\[2ex]
-\dfrac{\partial^2 L}{\partial R(1)\partial R(2)} & -\dfrac{\partial^2 L}{\partial[R(2)]^2} & -\dfrac{\partial^2 L}{\partial R(2)\partial R(3)} & \cdots & 0 & 0 \\[2ex]
0 & -\dfrac{\partial^2 L}{\partial R(2)\partial R(3)} & -\dfrac{\partial^2 L}{\partial[R(3)]^2} & \cdots & 0 & 0 \\[2ex]
\cdot & \cdot & \cdot & & \cdot & \cdot \\
\cdot & \cdot & \cdot & & \cdot & \cdot \\
0 & 0 & 0 & \cdots & -\dfrac{\partial^2 L}{\partial[R(M_2-1)]^2} & -\dfrac{\partial^2 L}{\partial R(M_2-1)\partial R(M_2)} \\[2ex]
0 & 0 & 0 & \cdots & -\dfrac{\partial^2 L}{\partial R(M_2-1)\partial R(M_2)} & -\dfrac{\partial^2 L}{\partial[R(M_2)]^2}
\end{bmatrix}^{-1}
$$

$$\tag{6.13}$$

Expressions for second partial derivatives can be obtained for Equation 6.7 and are available in Turnbull (1974). By minimizing the estimated variance of $R(t)$ in Equation 6.13, suitable values of a and b can be obtained. $Var[\hat{R}(t)]$ is used to develop confidence limits for $\hat{R}(t)$ ($t = 1, 2, 3, \dots, M_2$). The approximate 95% confidence limits for $\hat{R}(t)$ are thus obtained from the diagonal of inverse of the

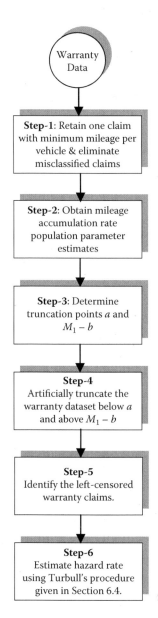

FIGURE 6.3   Flowchart of the six-step methodology.

Fisher's information matrix in Equation 6.13. The 95% confidence limits at each t are obtained as

$$\hat{R}(t) \pm 1.96\sqrt{Var[\hat{R}(t)]} \tag{6.14}$$

6.5 A SIX-STEP METHODOLOGY

A six-step methodology to arrive at the hazard rate estimate using the procedure explained in the previous section is shown as a flowchart in Figure 6.4.

Step 1: For every vehicle, warranty claims with minimum mileage are retained and repeat claims are screened out. The purpose is to keep manufacturing quality separate from service or repair quality. Misclassified and misdiagnosed claims are identified through one or more of the methods discussed in Chapter 3 and excluded from further analysis.

Step 2: This step deals with the issue of incompleteness of warranty data. Using the MIS and mileage values available for unfailed vehicles, estimates of population parameters for the mileage accumulation rates are obtained (steps 1 and 2 are the same as in Chapter 4, Section 4.3).

Step 3: Suitable values for the artificial truncation points a and $M_1 - b$ are determined using design of experiment methods by minimizing the standard deviation of maximum likelihood estimates of the reliability function as a response.

Step 4: Once the artificial truncation points a and $M_1 - b$ are determined in the previous step, the warranty data set is artificially truncated below a miles and above $M_1 - b$ miles.

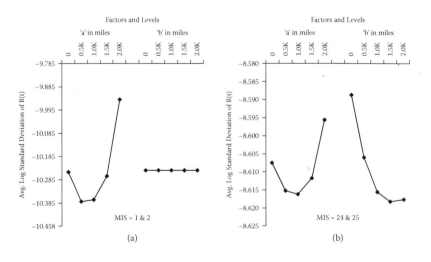

FIGURE 6.4 Average log standard deviation of reliability estimates at (a) MIS = 1 and 2, (b) MIS = 24 and 25.

Step 5: Using the methods suggested in Chapter 3, left-censored claims are identified for the first claims. When soft failures are such that the chance of detection by the user is low, one may consider all the claims between a and $M_1 - b$ miles as left censored. However, when the delayed reporting of failures is believed to have occurred only between $M_1 - b$ and M_1 miles, the search for left-censored claims may be restricted to this range only.

Step 6: The maximum likelihood estimates for the reliability function are obtained by iteratively using Equations 6.7–6.11. Subsequently, the estimates for cumulative hazard functions and their 95% confidence limits are obtained using Equations 6.12 and 6.14.

6.6   AN APPLICATION EXAMPLE

To illustrate the application of the methodology proposed in previous sections, we consider the field reliability analysis of a component-level failure mode using warranty data. The warranty coverage for the vehicle under study is two-dimensional with a time limit of 36 MIS and mileage limit of 36,000 mi, whichever occurs first. The number of vehicles in the field up to MIS $= t$ is denoted by V_t, as given in Table 6.1.

6.6.1   THE SIX STEPS

Step 1: There are in all 1002 first claims for the failure mode belonging to the soft failure category. After screening the claims, 48 claims were found to have incorrect failure mode attribution. These claims were excluded from further study.

Step 2: The estimates of location and shape parameters of the lognormal distribution are obtained as 7.28 and 0.689, respectively, as given in step 2 of Section 4.4 (Chapter 4).

Step 3: An experimental study is planned to arrive at suitable values of a and b. Five levels chosen for factors "a miles" and "b miles" are given in Table 6.2.

A full factorial experimental layout with 25 runs is used for the experiment. Standard deviation of maximum likelihood estimates of the reliability

TABLE 6.2
Factors and Levels Table for Determining the Truncation Points a and $M_1 - b$

Factor	Level 1	Level 2	Level 3	Level 4	Level 5
a miles	0	0.5K	1.0K	1.5K	2.0K
b miles	0	0.5K	1.0K	1.5K	2.0K

function is obtained at MIS = 1 and 2 using Equation 6.13. The stopping criteria used for the iterative process in Equations 6.7–6.11 is

$$\max_{1 < i < M_2} \left| R(i^l) - R(i^{l-1}) \right| < 10^{-15}$$

A logarithmic transformation of the standard deviation values is treated as the response from the experiment. The two response values obtained for MIS = 1 and MIS = 2 form the two replicates for the experiment. Since the aim is to minimize error in the estimation of the reliability function, the chosen response is of lower-the-better type. The analysis of the resulting data is performed using analysis of variance methodology. A similar analysis is carried out for responses obtained at MIS = 24 and 25. The results from the analysis of variance are summarized in Table 6.3.

From Table 6.3, it can be seen that for MIS = 1 and 2, effect of the factor a miles is statistically significant at the 0.04% level of significance. However, the effect of the factor b miles and the interaction effect between a miles and b miles are not found to be statistically significant on the response. Similarly, at MIS = 24 and 25, the main effects of factors a miles and b miles are found significant at <10% and 0.4% levels of significance, respectively. To enable choice of suitable levels for a and b, average response graphs are given for MIS = 1 and 2, and MIS = 24 and 25 in Figure 6.4a and Figure 6.4b, respectively.

From Figure 6.4a and Figure 6.4b, it is observed that for factor a miles, the average response initially decreases, reaching low values at level 2 and level 3, and then increases. Factor b miles does not have significant impact on the response, as seen in Figure 6.4a, because chances of vehicles

TABLE 6.3
Analysis of Variance for Determining the Truncation Points

Months in Service	Source of Variation	Sum of Squares	Degrees of Freedom	Mean Square	F-ratio	P
1 and 2	a	1.246	4	0.311	7.56	0.0004
	b	0.000	4	0.000	<1	
	Interaction	0.000	16	0.000	<1	
	Error	1.030	25	0.041		
	Total	2.276	49			
24 and 25	a	0.0028	4	0.0007	2.23	0.095
	b	0.0063	4	0.0016	5.00	0.004
	Interaction	0.0000	16	0.0000	<1	
	Error	0.0078	25	0.0003		
	Total	0.0169	49			

reaching 36K mi at 1 or 2 MIS are remote. Factor b miles in Figure 6.4b shows higher average response values at level 1 and level 2, and then gradually reaches low values at level 4 and level 5. Based on the experimental results, we choose the following:

$$a = 1000$$
$$b = 1500$$

Step 4: Using the values $a = 1000$ and $b = 1500$, the warranty data set is artificially truncated by excluding 38 claims below 1000 mi and 93 claims above 34,500 mi. Thus, out of the initial 1002 claims, 813 claims are used for further study.

Step 5: Using the detailed claim-related information available in the warranty database and engineering analysis of available field-returned parts, the 93 claims between 34,500 and 36,000 mi were individually analyzed to judge whether or not the actual failure might have occurred prior to the reported MIS. It was found that as many as 58 claims had potentially failed before the reported MIS. These 58 claims were regarded as left censored at the reported MIS values.

Step 6: Table 6.4 shows the initial and final versions of the iterative procedure. The cumulative hazard plot is shown in Figure 6.5 along with the 95% confidence limits.

From Figure 6.5, it is observed that the cumulative hazard plot shows a straight line up to about MIS = 30, indicating a constant failure rate trend. There seems to be a slight upward trend after MIS = 30, indicating an incipient increasing failure rate trend for the component-level failure mode under study.

6.6.2 COMMENTS ON RESULTS

The effect of the left artificial truncation point a and right artificial truncation point $M_1 - b$ as well as the inclusion of left-censored claims can be observed from the cumulative hazard plots depicted in Figure 6.6a–d.

Curve 1 in Figure 6.6a shows cumulative hazard rates under the assumption that all the failures reported within the mileage interval 1000 mi and 34,500 mi are the results of usage-related hard failures with a high chance of detection by users. Curve 2 considers 50% of the claims in the interval as belonging to the soft failure category, which had delayed failure reporting. It further assumes that all the failures within the warranty period are reported and result in warranty claims. The effect of considering certain percentage of claims as left censored is to adjust the effective number of reported failures at different MIS values. The left-censored claims lead to an upward correction in the number of first claims observed at MIS values prior to the reported MIS and a similar downward correction at MIS values higher than the MIS at which the claim was originally reported. A similar pattern can be observed in Figure 6.6b–d, although with different magnitudes at different MIS values. Thus, it is seen that treating the reported failures as hard or soft failures has a significant

TABLE 6.4

Reliability Estimates Using Turnbull's Iterative Procedure with Values at Iterations 0, 1, and at the Final Iteration 11

t	n_t	r_t	c_t	Probability of Mileage between 1K and 34.5K	Iteration 0			Iteration 1			Iteration 11		
					N_t	n_t	$R(t)$	N_t	n_t	$R(t)$	N_t	n_t	$R(t)$
1	13	0	0	0.705	141010.1	13	0.99991	141051.0	14.50	0.99990	141051.0	14.56	0.99990
2	20	0	0	0.939	187681.8	20	0.99980	187734.9	21.74	0.99978	187734.8	21.75	0.99978
3	19	0	0	0.982	196362.4	19	0.99970	196416.2	20.58	0.99968	196416.1	20.58	0.99968
4	37	0	0	0.990	197853.3	37	0.99952	197905.9	40.05	0.99947	197905.8	40.06	0.99947
5	36	0	0	0.986	197093.9	36	0.99933	197143.3	38.98	0.99928	197143.2	38.99	0.99928
6	39	0	0	0.976	195094.8	39	0.99914	195140.9	42.26	0.99906	195140.8	42.27	0.99906
7	35	0	0	0.962	192129.0	35	0.99895	192171.2	37.97	0.99886	192171.1	37.99	0.99886
8	27	0	1	0.943	188349.3	27	0.99881	188387.9	29.34	0.99871	188387.7	29.35	0.99871
9	33	0	1	0.921	183888.3	33	0.99863	183923.8	35.77	0.99851	183923.7	35.79	0.99851
10	40	0	0	0.896	178861.2	40	0.99841	178893.3	43.29	0.99827	178893.2	43.31	0.99827
11	36	0	0	0.869	173392.1	36	0.99820	173420.4	39.06	0.99805	173420.2	39.08	0.99805
12	46	0	5	0.840	167606.6	46	0.99793	167631.3	50.04	0.99775	167631.1	50.08	0.99775
13	33	0	0	0.810	161596.4	33	0.99772	161617.0	35.51	0.99753	161616.8	35.51	0.99753
14	38	0	2	0.779	155470.3	38	0.99748	155488.2	41.01	0.99727	155488.0	41.01	0.99727
15	37	0	4	0.748	149289.9	37	0.99723	149304.8	39.85	0.99700	149304.6	39.86	0.99700
16	34	0	2	0.718	143123.1	34	0.99699	143135.3	36.39	0.99675	143135.1	36.38	0.99675
17	25	0	2	0.687	137022.5	25	0.99681	137032.5	26.72	0.99655	137032.4	26.71	0.99655
18	29	0	1	0.657	131032.7	29	0.99659	131041.2	30.94	0.99632	131041.0	30.93	0.99632
19	19	0	6	0.628	125176.6	19	0.99644	125183.5	20.29	0.99616	125183.4	20.28	0.99615
20	20	0	1	0.599	119487.3	20	0.99627	119493.2	21.14	0.99598	119493.0	21.12	0.99598

21	27	0	5	0.572	113975.5	27	0.99604	113980.4	28.55	0.99573	113980.3	28.53	0.99573
22	17	0	1	0.545	108650.4	17	0.99588	108654.2	17.83	0.99557	108654.1	17.81	0.99557
23	15	0	3	0.520	103529.5	15	0.99574	103532.7	15.73	0.99542	103532.6	15.71	0.99541
24	13	0	1	0.495	98612.5	13	0.99561	98615.2	13.57	0.99528	98615.2	13.56	0.99528
25	9	20000	5	0.471	93901.1	9	0.99551	93903.4	9.39	0.99518	93903.4	9.38	0.99518
26	22	20000	2	0.449	80421.9	22	0.99524	80424.0	22.82	0.99490	80423.9	22.80	0.99490
27	19	20000	2	0.427	68001.8	19	0.99496	68003.4	19.72	0.99461	68003.4	19.70	0.99461
28	13	20000	0	0.406	56585.0	13	0.99473	56586.2	13.50	0.99437	56586.2	13.49	0.99437
29	4	20000	5	0.387	46110.0	4	0.99465	46110.9	4.19	0.99428	46110.9	4.19	0.99428
30	9	20000	0	0.368	36516.9	9	0.99440	36517.8	9.31	0.99403	36517.8	9.30	0.99403
31	11	20000	1	0.350	27741.7	11	0.99401	27742.4	11.50	0.99361	27742.4	11.49	0.99362
32	7	15000	3	0.333	19729.4	7	0.99365	19729.9	7.38	0.99324	19729.9	7.38	0.99324
33	5	15000	2	0.317	14015.1	5	0.99330	14015.5	5.22	0.99287	14015.5	5.22	0.99287
34	6	10000	1	0.302	8807.5	6	0.99262	8807.8	6.22	0.99217	8807.8	6.21	0.99217
35	14	10000	1	0.288	5506.9	14	0.99010	5507.1	14.46	0.98957	5507.1	14.45	0.98957
36	6	9129	1	0.274	2501.2	6	0.98772	2501.2	6.19	0.98712	2501.2	6.19	0.98712

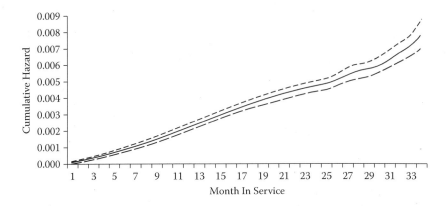

FIGURE 6.5 Cumulative hazard rate.

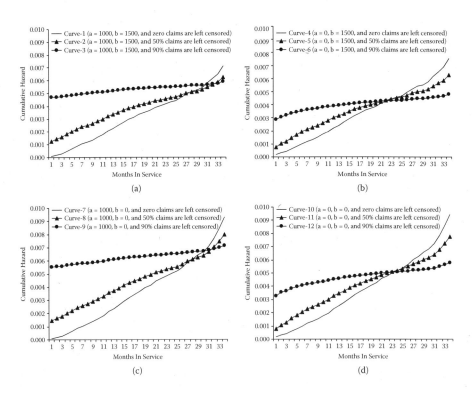

FIGURE 6.6 Comparison of cumulative hazard rate estimates at different artificial truncation points, with different percentages of left-censored claims.

impact on the cumulative hazard rate pattern irrespective of whether the left or right artificial truncation points are considered or not.

Figure 6.7 shows the curves 1, 4, 7, and 10 from Figure 6.6 and compares the cumulative hazard plots for situations when left or right or both truncation points are considered and the failure mode belongs to the hard failure category. Curve 1 in Figure 6.7 addresses both spikes at the beginning and toward the end of the warranty coverage period while arriving at the cumulative hazard rate. For curve 4, only the spike in claims due to customer-rush near the warranty expiration limit is addressed. The existence of manufacturing or assembly defects in addition to usage-related failures lead to slightly higher cumulative hazard rates in curve 4. Curve 7 ignores the phenomena of customer-rush and as a result shows higher cumulative hazard rates beyond about MIS = 20 as compared to the curves 1 and 4. Below MIS = 15, curves 1, 4, 7, and 10 show a relatively small difference in the cumulative hazard rates, as not a significant portion of the vehicle population is expected to exceed 36K mi by then. From this figure, it is seen that the impact of ignoring spikes in warranty claims toward the end of warranty coverage is more pronounced than at the beginning. This is due to a larger accumulation of claims as a result of the customer-rush phenomena. However, when the warranty data shows a higher spike toward the beginning of the warranty period, the impact of changing the left artificial truncation point, denoted by a, would be more pronounced.

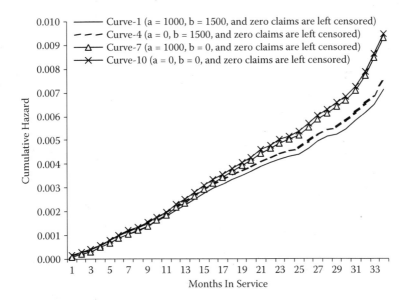

FIGURE 6.7 Comparison of cumulative hazard rate estimates with different artificial truncation points and no left censoring.

6.7   SUMMARY

Key points from the chapter are summarized as follows:

- Research in the area of field reliability studies using warranty data has traditionally assumed reported time or mileage at failure in warranty claims to be the actual time or mileage at failure. Although this is largely true for hard failures, there may be delayed reporting of soft failures. In certain situations, reporting can be delayed to such an extent that failures get reported when the warranty coverage is about to expire. Such situations lead to the phenomena of customer-rush near the warranty expiration limit, resulting in an unusually high number of claims toward the end of the warranty coverage period. Ignoring such phenomena may cause the quality and reliability engineers to incorrectly obtain an increasing failure rate pattern from the cumulative hazard plots during field reliability studies. This in turn may lead to costly and uncalled-for component or subsystem design changes. It is thus critical to distinguish between warranty claims belonging to soft or hard failure categories and use suitable methods for field reliability studies.
- This chapter presents a methodology that helps to obtain nonparametric estimates of cumulative hazard rates when the warranty claims may potentially be a result of soft failures. The methodology requires that estimates of mileage accumulation rates in the vehicle population be available. The methodology proposed also helps to obtain approximate confidence limits for the maximum likelihood estimates of the cumulative hazard function.
- The chapter also provides a method to arrive at artificial truncation points a and $M_1 - b$ using the approach of experimental design.

BIBLIOGRAPHIC NOTES

In reliability studies using automobile warranty data, the time or mileage at the time of a warranty claim is assumed to be the time or mileage at failure (Suzuki 1985a, 1985b). Similarly, an absence of warranty claims is treated as a "no failure" situation. Kalbfleisch, Lawless, and Robinson (1991); Lawless, Hu, and Cao (1995); Lu (1998); Kalbfleisch and Lawless (1992); Stephens and Crowder (2004); and Rai and Singh (2003a, 2003b, 2004b) also provide warranty data analysis methods with examples based on such an assumption. Iskandar and Blischke (2003), while carrying out reliability analysis using warranty data from motorcycle claims, point out delay in the reporting of noncritical failures. They suggest that such delays in reporting may be modeled by fitting a mixture of distributions to the usage data. They, however, do not pursue such an analysis in the paper.

7 Soft Failures with Unknown Mileage Accumulation Rates*

There are things known and there are things unknown, and in between are the doors of perception.

—Aldous Huxley

OBJECTIVES

This chapter explains the following:

- Hazard function for soft failures when mileage accumulation rates in the vehicle population are not known
- Methodology to estimate population parameters of a doubly truncated normal or lognormal distribution
- Iterative procedure to incorporate censoring information when estimating hazard rates
- A seven-step methodology to estimate the hazard rate, with an example

OVERVIEW

This chapter provides a methodology to arrive at hazard rate estimates for failure modes belonging to the soft failure category. Section 7.1 gives the hazard function for situations when the warranty data set is doubly truncated and the mileage accumulation rate in the vehicle population is not known. Section 7.2 provides a methodology to estimate the population parameters of a doubly truncated data set known to follow a lognormal or normal distribution, which in turn helps to estimate the total number of first claims at each MIS (months in service). Section 7.3 provides an iterative procedure that helps to arrive at the maximum likelihood estimates of the hazard function while incorporating the censoring information associated with the warranty data. Section 7.4 gives a flowchart of the steps involved in arriving at the hazard function, and Section 7.5 illustrates the use of the methodology with an application example.

---

* © 2004. Elsevier. Reprinted with permission from Elsevier: Rai, B. K. and Singh, N. (2004). Modeling and analysis of automobile warranty data in presence of bias due to customer-rush near warranty expiration limit. Reliability Engineering and System Safety, 86, 83–94.

7.1   HAZARD FUNCTION

The mileage accumulation rate for the vehicle population under study sometimes may not be available to the engineers carrying out the field reliability study. This situation calls for making adjustments to the numerator of the hazard function (Chapter 3, Equation 3.1). When the "spikes" in the number of claims occur below a miles and above $M_1 - b$ miles, the warranty data set is artificially truncated below and above these points, respectively. After a modification to the numerator of Equation 3.1, we obtain

$$h_6(t) = \frac{n_{L,t} + n_t + n_{R,t}}{N(t)} = \frac{n(t)}{N(t)} \tag{7.1}$$

where
  n_t: Number of first claims between a and $M_1 - b$ miles at MIS $= t$
  $n_{L,t}$: Number of first failures below a miles at MIS $= t$
  $n_{R,t}$: Number of first failures above $M_1 - b$ miles at MIS $= t$
  $n(t)$: Total number of first failures expected at MIS $= t$

Estimates of $n_{L,t}$ and $n_{U,t}$ can be obtained by treating the data on first claims to be left truncated at a miles and right truncated at $(M_1 - b)$ miles, and then estimating the parameters of the complete distribution. This process is repeated at each MIS value. We consider cases where mileage to first claim is expected to follow a lognormal or normal distribution. Figure 7.1 shows truncated distributions at different MIS values.

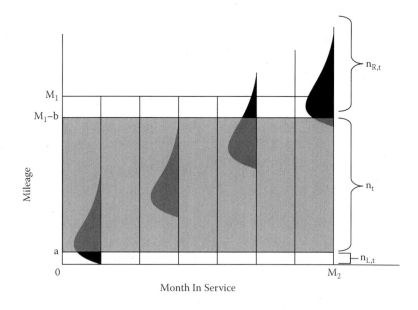

FIGURE 7.1   Truncated distributions of mileage to first failure.

Figure 7.1 shows different possible scenarios with real data. Note that a constant underlying distribution is assumed at each MIS value. However, the distributions may have different shape and scale parameters. The distributions toward lower values of MIS are essentially left truncated at a miles with substantial probabilities below a miles when compared to similar values toward higher MIS. It is thus expected that the estimated values of $n_{L,t}$ will be higher toward lower MIS values and almost negligible toward higher MIS values. However, the opposite is true for the region beyond the right truncation point of $M_I - b$ miles.

7.2 ESTIMATION FROM DOUBLY TRUNCATED DATA SETS

This section provides a methodology to arrive at the estimates of population parameters from doubly truncated data sets. The scope of such estimation is limited to the normal and lognormal distributions. Once estimates of population parameters for doubly truncated data sets are arrived at each MIS, the total number of first failures at each MIS can be easily estimated.

Let Y_t be a random variable denoting the number of miles traveled by failed vehicles at MIS $= t$. When the lognormal distribution provides a good approximation to the distribution of mileage on failed vehicles at a given MIS value, the logarithmic transformation $X = \ln(Y)$ representing normal distribution is easier to deal with.

Let a and $M_I - b$ be the left and right truncation points for mileage on failed vehicles such that $a < (M_I - b)$. Let $\ln(a) = X_L$ and $\ln(M_I - b) = X_R$ be the left and right truncation points, respectively, in terms of the normal variable. Let the number of first claims between the two truncation points at an MIS $= t$ be denoted by n_t. If an observation x_i is not included in the data set unless it lies between two known points of truncation, that is, $X_L \leq x_i \leq X_R$, it is said to be truncated on the left at X_L and on the right at X_R. When the mean

$$\left(\overline{x_t} = \sum_{i=1}^{n_t} x_i / n_t \right)$$

is calculated from such doubly truncated data, it is either an underestimate (toward higher values of MIS) or an overestimate (toward lower values of MIS) of the population mean μ. When the variance

$$\left(S_t^2 = \sum_{i=1}^{n_t} (x_i - \overline{x})^2 / [n_t - 1] \right)$$

is calculated from such doubly truncated data, it is an underestimate of the population variance σ^2.

From estimates of population parameters μ and σ, estimated total number of first failures, say $n(t)$, at any MIS value is obtained as

$$\hat{n}(t) = \frac{n_t}{P(X_L < X < X_R)} \tag{7.2}$$

For a doubly truncated normal distribution with truncation points at $x = X_L$ and $x = X_R$, the pdf can be written as

$$f_T(x;\mu,\sigma) = \frac{\dfrac{1}{\sigma\sqrt{2\pi}} e^{-\left|\frac{1}{2}\left(\frac{x-\mu}{\sigma}\right)^2\right|}}{\dfrac{1}{\sigma\sqrt{2\pi}} \displaystyle\int_{X_L}^{X_R} e^{-\left|\frac{1}{2}\left(\frac{x-\mu}{\sigma}\right)^2\right|} dx}$$

$$= \frac{1}{[F(X_R) - F(X_L)]\sigma\sqrt{2\pi}} \exp\left[-\frac{1}{2}\left(\frac{x-\mu}{\sigma}\right)^2\right]$$

$$= \frac{1}{[\Phi(\xi_R) - \Phi(\xi_L)]\sigma\sqrt{2\pi}} \exp\left[-\frac{1}{2}\left(\frac{x-\mu}{\sigma}\right)^2\right] \qquad (X_L < x \le X_R)$$

$$= 0 \text{ elsewhere} \tag{7.3}$$

where

$$\Phi(\xi_L) = \frac{1}{\sqrt{2\pi}} \int_{-\infty}^{\xi_L} e^{-\frac{t^2}{2}} dt = F(X_L) = \frac{1}{\sqrt{2\pi}\sigma} \int_{-\infty}^{X_L} e^{-\frac{1}{2}\left(\frac{x-\mu}{\sigma}\right)^2} dt$$

$$\Phi(\xi_R) = \frac{1}{\sqrt{2\pi}} \int_{-\infty}^{\xi_R} e^{-\frac{t^2}{2}} dt = F(X_R) = \frac{1}{\sqrt{2\pi}\sigma} \int_{-\infty}^{X_R} e^{-\frac{1}{2}\left(\frac{x-\mu}{\sigma}\right)^2} dt$$

$$\xi_L = \frac{X_L - \mu}{\sigma} \; ; \; \xi_R = \frac{X_R - \mu}{\sigma}$$

Let $x_1, x_2, ..., x_{n_t}$ be a random sample of size n_t from a population with pdf given by Equation 7.3. Then, the loglikelihood function of the sample values $x_1, x_2, ..., x_{n_t}$ is given by

$$\ln L = \ln\left[f_T(x_1;\mu,\sigma) f_T(x_2;\mu,\sigma)...f_T(x_{n_t};\mu,\sigma) \right]$$

$$= \sum_{i=1}^{n_t} f_T(x_i;\mu,\sigma)$$

$$= -n_t \ln\left[\Phi(\xi_R) - \Phi(\xi_L) \right] - \frac{1}{2\sigma^2} \sum_{i=1}^{n_t} (x_i - \mu)^2 - n_t \ln\sigma + const. \tag{7.4}$$

Maximum likelihood estimators for μ and σ are determined by obtaining partial derivatives of Equation 7.4 with respect to μ and σ and equating them to zero.

7.2.1 Use of a Special Chart and Table to Obtain MLEs

Cohen (1957) derived the likelihood equations from Equation 7.4 in the following form:

$$\frac{\partial \ln L}{\partial \mu} = \frac{\bar{Z}_1 - \bar{Z}_2 - \xi_L}{\xi_R - \xi_L} - \frac{v_1}{w} = H_1(\xi_L,\xi_R) - \frac{v_1}{w} = 0 \tag{7.5}$$

$$\frac{\partial \ln L}{\partial \sigma} = \frac{1 + \xi_L \bar{Z}_1 - \xi_R \bar{Z}_2 - (\bar{Z}_1 - \bar{Z}_2)^2}{(\xi_R - \xi_L)^2} - \frac{s^2}{w^2} = H_1(\xi_L,\xi_R) - \frac{s^2}{w^2} = 0 \tag{7.6}$$

where

$$\bar{Z}_1 = \frac{\phi(\xi_L)}{\Phi(\xi_R) - \Phi(\xi_L)} \; ; \; \bar{Z}_2 = \frac{\phi(\xi_R)}{\Phi(\xi_R) - \Phi(\xi_L)} \; ; \; w = X_R - X_L \; ; \; v_1 = \bar{x} - X_L$$

By simultaneously solving Equations 7.5 and 7.6, estimates of $\hat{\xi}_L$ and $\hat{\xi}_R$ can be obtained. Estimates of μ and σ can then be obtained as

$$\hat{\sigma} = \frac{w}{\hat{\xi}_R - \hat{\xi}_L} \quad \text{and} \quad \hat{\mu} = X_L - \hat{\sigma}\hat{\xi}_L \tag{7.7}$$

To simplify computations for the estimation of $\hat{\xi}_L$ and $\hat{\xi}_R$, Cohen (1957, 1991) provided a special chart of intersecting graphs of equations v_1/w and s^2/w^2, and a table that can be used for interpolation when improved accuracy of estimates is needed.

The use of the special chart and the table to estimate the total number of first failures is illustrated using an example.

EXAMPLE 7.1

Consider the mileage data for a certain failure mode of a vehicle part that is known to follow a lognormal distribution. The data are truncated below $a = 1K$ mi and above $M_I - b = 35K$ mi at MIS = 3. Totally, 19 claims were recorded between this band with following mileage values:

Y (in miles): 1579, 1791, 2024, 2111, 2435, 2476, 2947, 2992, 3498, 4549, 5235, 5909, 6880, 10110, 10964, 11045, 11510, 21074, 25364

Using these $n_t = 19$ observations, we need an estimated number of total first failures at MIS = 3. As the data follows the lognormal distribution, logarithmic transformation of the preceding data gives

$X = \ln(Y)$: 7.3645, 7.4905, 7.6128, 7.6549, 7.7977, 7.8144, 7.9885, 8.0037, 8.1599, 8.4227, 8.5631, 8.6842, 8.8364, 9.2213, 9.3024, 9.3097, 9.3510, 9.9558, 10.1411

Also,

$X_L = \ln(1000) = 6.908$; $X_R = \ln(35000) = 10.463$; $\bar{x} = 8.5092$; $s = 0.8468$;

$w = X_R - X_L = 3.555$; $v_1 = \bar{x} - X_L = 1.6014$; $v_1 / w = 0.4504$; $s^2 / w^2 = 0.0567$

Step 1: Obtain the initial estimates of $\hat{\xi}_L$ and $\hat{\xi}_R$ from the chart given by Cohen (1957) from the intersection of lines corresponding to $v_1 / ws = 0.4504$ and $s^2 / w^2 = 0.0567$. We have

$$\hat{\xi}_L^0 = -1.3 \text{ and } \hat{\xi}_R^0 = 1.9$$

Step 2: Using the table given by Cohen (1991), obtain four data points in the neighborhood of H_1 and H_2 as

ξ_L	ξ_R	H_1	H_2
−1.3	1.9	0.444041	0.057427
−1.3	1.8	0.453730	0.059096
−1.4	1.9	0.452864	0.056538
−1.4	2.0	0.443174	0.054873

Step 3: Using ξ_L, ξ_R, H_1, and H_2, obtain the following two multiple regression equations:

$$\xi_R = a_1 + b_1 H_1 + c_1 \xi_L$$

$$\xi_R = a_2 + b_2 H_2 + c_2 \xi_L$$

The two equations obtained for this example are

$$\xi_R = 5.29896 - 10.32045 H_1 - 0.91057 \xi_L$$

$$\xi_R = 6.03826 - 59.98792 H_2 + 0.53329 \xi_L$$

Step 4: From the two equations, obtain two points each on curve H_1 and H_2.

Points on H_1 : (−1.3, 1.83437) and (−1.4, 1.92543)

Points on H_2 : (−1.3, 1.94367) and (−1.4, 1.89034)

Step 5: Obtain equations of straight lines either from the slope and intercept using points on H_1 and H_2 obtained in the previous step or directly by replacing values $H_1 = v_1/w = 0.4504$ and $H_2 = s^2/w^2 = 0.0567$ in the two equations obtained in step 3. Thus, we have

$$0.91057 \xi_L + \xi_R = 0.65063$$

$$-0.53329 \xi_L + \xi_R = 2.63695$$

Step 6: From the simultaneous solution of the two linear equations, obtain $\hat{\xi}_L$ and $\hat{\xi}_R$. The two equations of the straight line obtained in step 5 can be written in matrix form as

$$\begin{bmatrix} 0.91057 & 1 \\ -0.53329 & 1 \end{bmatrix} \begin{bmatrix} \xi_L \\ \xi_R \end{bmatrix} = \begin{bmatrix} 0.65063 \\ 2.63695 \end{bmatrix} \Rightarrow \begin{bmatrix} \hat{\xi}_L \\ \hat{\xi}_R \end{bmatrix} = \begin{bmatrix} -1.3757 \\ 1.9033 \end{bmatrix}$$

Step 7: From the estimates of $\hat{\xi}_L$ and $\hat{\xi}_R$, we obtain

$$\hat{\sigma} = \frac{w}{\hat{\xi}_R - \hat{\xi}_L} = 1.08417 \quad \text{and} \quad \hat{\mu} = X_L - \hat{\sigma}\hat{\xi}_L = 8.3995$$

Step 8: Estimate $n(t)$ using Equation 7.2:

$$\hat{n}(t) = \frac{n_t}{P(X_L < X < X_R)} = 21$$

Thus, using the data on 19 first claims between 1K and 35K mi, it is estimated that there are totally 21 first failures at MIS = 3.

7.2.2   MAXIMUM LIKELIHOOD ESTIMATE USING AN ITERATIVE PROCEDURE

From Equations 7.5 and 7.6, a quadratic expression of the following form can be derived:

$$at^2 + bt + c = 0 \tag{7.8}$$

By applying the quadratic formula, we can obtain the following solution:

$$t = \frac{-b \pm \sqrt{b^2 - 4ac}}{2a} \tag{7.9}$$

From Equations 7.5 and 7.6, Cohen (1957) developed a quadratic expression of the form shown in Equation 7.8 with $t = \xi_L - \xi_R$, and using Equation 7.9 obtained following expressions for the iterative procedure:

$$\xi_R^{(i+1)} = \left(\overline{Z}_1^{(i)} + \overline{Z}_2^{(i)}\right) +$$

$$\frac{\left\{\left[\left(\overline{Z}_1^{(i)} - \overline{Z}_2^{(i)}\right)\frac{v_1}{w} + \overline{Z}_2^{(i)}\right] - \sqrt{\left[\left(\overline{Z}_1^{(i)} - \overline{Z}_2^{(i)}\right)\frac{v_1}{w} + \overline{Z}_2^{(i)}\right]^2 + \frac{4s^2}{w^2}}\right\}w(v_1 - w)}{2s^2} \tag{7.10}$$

$$\xi_L^{(i+1)} = \frac{-\left[\xi_R^{(i+1)}\frac{v_1}{w} - \left(\overline{Z}_1^{(i)} - \overline{Z}_2^{(i)}\right)\right]}{1 - \frac{v_1}{w}} \tag{7.11}$$

Although one can obtain initial estimates of ξ_L and ξ_R from the special chart, Equations 7.10 and 7.11 can be used independently to obtain the required estimates. Working with various examples, the authors have observed that even when relatively distant starting points such as $\xi_L^{(0)} = -10$ and $\xi_R^{(0)} = 10$ are used, accurate estimates of ξ_L and ξ_R can be obtained rather quickly using simple computer programs.

7.2.3   COVARIANCE MATRIX OF THE MAXIMUM LIKELIHOOD ESTIMATOR

An approximate variance–covariance matrix of $\hat{\mu}$ and $\hat{\sigma}$ is obtained as an inverse of Fisher's information matrix, that is,

$$\begin{bmatrix} Var(\hat{\mu}) & Cov(\hat{\mu},\hat{\sigma}) \\ Cov(\hat{\mu},\hat{\sigma}) & Var(\hat{\sigma}) \end{bmatrix} = \begin{bmatrix} -\dfrac{\partial^2 \ln L}{\partial \mu^2} & -\dfrac{\partial^2 \ln L}{\partial \mu \partial \sigma} \\ -\dfrac{\partial^2 \ln L}{\partial \mu \, \partial \sigma} & -\dfrac{\partial^2 \ln L}{\partial \sigma^2} \end{bmatrix}^{-1} \tag{7.12}$$

Expressions for second partial derivatives can be obtained from Equations 7.5 and 7.6, and have been given by Cohen (1957; 1991).

7.3   INCORPORATING CENSORING INFORMATION IN THE HAZARD FUNCTION ESTIMATION

From the methods discussed in the previous section, we can obtain an estimate of the total number of first failures at each MIS. However, to obtain the hazard function, we need to adjust the numerator in Equation 7.1 to account for left-censored cases that can exist in case of soft failures. To make such an adjustment and enable estimation of the appropriate hazard rate, we make use of the iterative procedure developed by Turnbull (1974). The iterative procedure discussed in Section 6.4 of Chapter 6 is used to arrive at the cumulative hazard rate except that the probability term in Equation 6.9 is taken out. For ease of reference, the remaining equations are repeated:

$$R(t) = q_t \times R(t-1), \ t = 2, 3, \ldots, M_2 \tag{7.13}$$

With, $R(0) = 1$; $R(1) = q_1$, and

$$q_t = \frac{\left(N'(t) - n_t'\right)}{N'(t)} \tag{7.14}$$

$$N'(t) = \sum_{i=t}^{M_2} (r_i + n_i') \tag{7.15}$$

$$n_t' = n_t + \sum_{i=t}^{M_2} c_i \alpha_{ij} \qquad (7.16)$$

$$\alpha_{it} = \frac{R(t-1) - R(t)}{1 - R(i)}, t \leq i \qquad (7.17)$$

The iteration procedure, as discussed before, starts with obtaining $R(t_i^0)$ as initial estimates assuming $c_i = 0, \forall i = 1,2,...,m$. Then using Equations 7.13–7.17, values for $R(t_i^l), l \geq 0$ are iteratively obtained until $\max_{1<i<m} |R(t_i^l) - R(t_i^{l-1})| < \delta$, where δ is a very small number.

From the estimates $\hat{R}(t_i)$, the cumulative hazard function $H(t)$ is obtained from the following relationship:

$$H(t) = -\ln[\,\hat{R}(t_i)\,] \qquad (7.18)$$

7.4 A SEVEN-STEP METHODOLOGY

A seven-step methodology to arrive at the hazard rate estimate based on the procedure described in the previous section is shown in the form of a flowchart in Figure 7.2.

Step 1: Same as the step explained in previous chapters.

Step 2: The artificial truncation points a and $M_1 - b$ are determined (as discussed in Chapter 6) by using design of experiment methods, by minimizing the standard deviation of maximum likelihood estimates of the reliability function as a response. Another simple way to identify these points is by looking at the histogram of the number of vehicles at different MIS values. Once the truncation points a and $M_1 - b$ are determined, the warranty data set can be artificially truncated below a miles and above $M_1 - b$ miles. This action makes the warranty data at each MIS value doubly truncated.

Step 3: Using Equations 7.10 and 7.11, the population parameter estimates for the doubly truncated warranty data set is arrived at for each MIS.

Step 4: In this step, unusual observations that are outside the 99% statistical limits are identified and population parameters are reestimated with the remaining data. This ensures that stabilized estimates for population parameters are obtained.

Step 5: The total number of first failures at each MIS is obtained using Equation 7.2.

Step 6: The expected number of left-censored claims in the interval $M_1 - b$ and M_1 can also be obtained using the population parameter estimates. This is obtained by subtracting the expected number of first claims in the interval $M_1 - b$ and M_1 from the observed number of first claims at each MIS.

Step 7: The cumulative hazard rates are obtained using Equations 7.13–7.18.

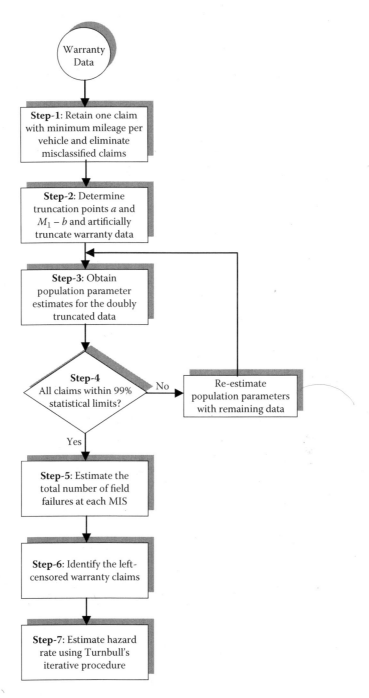

FIGURE 7.2   Flowchart of the seven-step methodology for estimation of the hazard function from soft failures with unknown mileage accumulation rates in the vehicle population.

7.5 AN APPLICATION EXAMPLE

Consider an automobile warranty data set of a previous model-year vehicle with a population of $N = 200,000$. Each vehicle in the population is covered by a two-dimensional warranty of 3 years or 36K mi, whichever occurs first. The mileage distribution of first claims at each MIS is assumed to follow the lognormal distribution. To work with the normal distribution, the data on mileages are simply subjected to logarithmic transformation. The number of vehicles in the field at each MIS is given in Table 6.1.

> **Step 1:** *Data quality.* This step is the same as that discussed in Chapter 6, Section 6.5.
>
> **Step 2:** *Artificial truncation of warranty data.* The histogram of a number of vehicles at different MIS for the failure mode under study is shown as failure mode 1 and vehicle A in Figure 6.1. By observing the figure, we select $a = 1,000$ mi and $M_1 - b = 35,000$ mi. The warranty data are artificially truncated below 1000 mi and above 35,000 mi. Consider an example of warranty data at MIS = 15. There are 46 vehicles with first claims at MIS = 15. The mileage values on these 46 vehicles range from a minimum of 7,595 mi to a maximum of 35,945 mi. Thus, none of the vehicles have a mileage below $a = 1000$ mi. There are 5 vehicles with mileages greater than 35K mi, and some of these claims may be potentially influenced by customer-rush near the warranty expiration limit. These 5 claims were removed from further analysis to avoid bias in the estimates. The mileage distribution of the remaining 41 claims is shown in Figure 7.3.

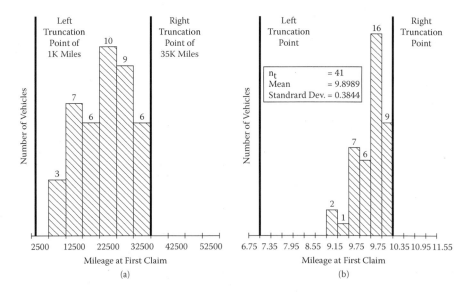

FIGURE 7.3 Histogram of the number of vehicles with first claims at MIS = 15: (a) mileage distribution, (b) mileage distribution after logarithmic transformation.

Figure 7.3a shows the mileage distribution of 41 claims between 1K mi and 35K mi. Figure 7.3b shows a similar distribution after applying logarithmic transformation to the mileage values. This figure indicates that at MIS = 15, the probability of a vehicle failure below the left truncation point of 1K mi is expected to be almost negligible. At the same time, vehicle failures are likely to occur beyond the right truncation point. Hence, the mean, 9.8989, obtained from the truncated data set of 41 claims is likely to be an underestimate of the mean of a complete distribution expected at MIS = 15. Similarly, the standard deviation of 0.3844 is also likely to be an underestimate of the standard deviation of the complete distribution.

Steps 3 and 4: *Iterative procedure.* Parameters for the parent or complete distribution are obtained through an iterative procedure using Equations 7.10 and 7.11. A starting point of $\xi_L^{(0)} = -10$ and $\xi_R^{(0)} = 10$ is used to obtain estimates at each MIS. Figure 7.4 shows the graphs of estimated standard truncation points during the iteration process for various MIS values.

It can be seen from Figure 7.4 that for MIS = 2, 3, and 5, the convergence to the final estimate is relatively quicker as compared to that at MIS = 1, 10, 15, 20, 25, and 30. However, even on an ordinary computer, the difference between the former and latter in obtaining the final estimates is no more

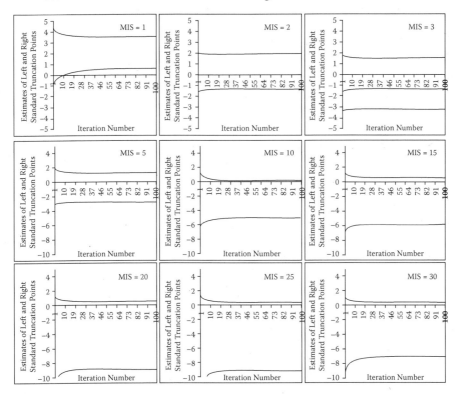

FIGURE 7.4 Estimation of left and right standard truncation points for selected MIS values.

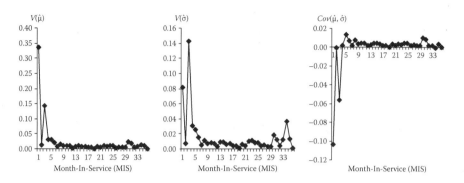

FIGURE 7.5   Variance and covariance of the estimates.

than a few seconds. The chart for MIS = 1 indicates that the data set is more left truncated than right truncated. This is expected, as not many first claims are likely to have a mileage over 35K mi at this stage. On the other hand, at MIS = 30, we observe more of right truncation than left truncation. At this stage, almost all vehicles in the parent distribution are expected to have a mileage greater than 1000 mi.

   Variance and covariance of the estimates. The reliability of estimates is studied by obtaining the variance and covariance of the estimates using Equation 7.12. Plots of $V(\hat{\mu})$, $V(\hat{\sigma})$, and $Cov(\hat{\mu},\hat{\sigma})$ at each MIS value are shown in Figure 7.5.

   It is observed that variance of the estimates for the first few MIS values are high in all the plots. One of the likely reasons is the existence of a combination of reliability and quality issues beyond 1000 mi for the initial MIS values (Rai and Singh 2003a, 2003b).

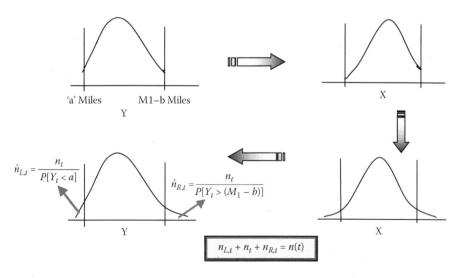

FIGURE 7.6   Estimating total number of failures.

Step 5: *Estimate of total number of first failures.* From the estimates of $\hat{\mu}$ and $\hat{\sigma}$, the total number of first failures at each MIS value is obtained using Equation 7.2, as shown in Figure 7.6. For example, the estimates of $\hat{\mu}$ and $\hat{\sigma}$ at MIS = 15 are 10.15613 and 0.54119, respectively. Using Equation 7.2 the total number of claims at MIS = 15 is obtained as

$$\hat{n}(t) = \frac{n_t}{P(X_L < X < X_R)} = \frac{41}{0.7147} \cong 57$$

Note that we use the term *failures* and not *claims* in the phrase *estimate of total number of first failures.* This is because the total number of first failures includes the estimated first failures beyond the warranty mileage limit that do not become a warranty claim.

Step 6: *Number of first claims expected to be left censored at each MIS.* The number of first claims expected to be left censored is obtained by subtracting the expected number of first claims between 35,000 and 36,000 mi from the observed number of first claims at each MIS value. For example, using the estimates of $\hat{\mu}$ and $\hat{\sigma}$ at MIS = 15, we obtain the probability of failure during 35K mi and 36K mi as

$$P\left[\ln(35000) < X < \ln(36000)\right] = 0.0174$$

From these estimates, one would expect $0.0174 \times 57 = 0.99 \cong 1$ claim, with the reported mileage between 35K mi and 36K mi at MIS = 15. Thus:

Number of first claims expected to be left censored at MIS = 15
= (Observed number of 1st claims at MIS = 15) −
(Expected number of 1st claims within the last 1K mi of warranty coverage at MIS = 15)
= 5 − 1
= 4

We treat these additional four claims as being contributed by customer-rush phenomena and representing the left-censored case.

Step 7: *Hazard plot.* In addition to the left-censored data arrived at in the previous step, there exist right-censored vehicles in Table 6.1 that are in the field and have not yet failed. To obtain the cumulative hazard plot from such doubly censored data, the iterative procedure explained in Section 7.3 is used.

Comparison of cumulative hazard plots with and without addressing accumulation of claims near warranty mileage limit. For comparison purposes, the cumulative hazard plot (see Step 7) was also arrived at by ignoring the accumulation

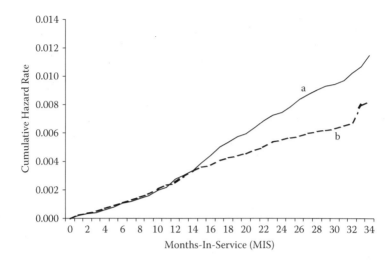

FIGURE 7.7 Cumulative hazard plot for two cases by treating warranty data as (a) only right truncated at 36K mi, (b) doubly truncated and doubly censored.

of first claims near the beginning and end of the warranty mileage limit, by treating data only as right truncated at 36K mi. Figure 7.7 shows the cumulative hazard plots for both cases.

Curve a in Figure 7.7 suggests an increasing failure rate (IFR) from MIS = 1 to about MIS = 30. At the same time, the shape of curve b suggests more of a constant failure rate (CFR). Up to about MIS = 15, the pattern for the cumulative hazard rate is similar for both situations. The difference in the two curves is more prominent beyond MIS = 15. A rough estimate of the mileage accumulation rate for the vehicle line under study indicated that about 25% of the vehicle population had completed 36K mi by MIS = 15. By MIS = 25, as many as 50% of the vehicle population had driven beyond the warranty mileage limit of 36K mi. This indicates that the accumulation of first claims close to the warranty mileage limit is expected to gradually increase beyond MIS = 15. Figure 7.7 also indicates that the accumulation of first claims near the start of the warranty period does not impact the overall pattern of the cumulative hazard plot in a noticeable way. However, this observation is limited to the current failure mode under study.

Since curve b addresses the phenomena of accumulation of first claims near the warranty mileage limit, it presents a more representative feedback on field failures than curve a, which does not address such phenomena. The different nature of design- or manufacturing-related improvement actions apply when an IFR or CFR pattern is observed from the cumulative hazard plot. Thus, methods proposed in this chapter to address customer behavior toward soft failures help to provide design engineers with a more representative feedback of field failures and enable improvements in current and forward model-year vehicles.

7.6   SUMMARY

The key points from the chapter are summarized as follows:

- Soft failures that have a high chance of detection by users sometimes may get reported late. In extreme cases the delay could result in accumulation of a large number of such claims near the warranty expiration limit. Such a phenomenon needs to be captured proactively at the time of modeling and analysis of warranty data to avoid obtaining a distorted picture of field failures.
- When mileage accumulation rates in the vehicle population are not available, the calculations to arrive at hazard rate estimates are more involved. Artificial truncation to avoid bias in the warranty data set toward the beginning and end of the warranty period makes the data set doubly truncated. The estimation of the population parameters help to arrive at failures expected outside the truncation points. The methodology also helps to estimate the expected number of left censored claims near the warranty expiration limit.
- This chapter has provided a detailed methodology, with illustrative examples, to arrive at hazard rate estimates in the presence of customer-rush near the warranty expiration limit, as well as for soft failures in general.

Section III

Warranty Prediction

8 Assessing the Impact of New Time/Mileage Warranty Limits*

Great is the art of beginning, but greater is the art of ending.

—Lazurus Long

OBJECTIVES

This chapter explains the following:

- A simple method to estimate the number and cost of claims per 1000 vehicles in the field at new time or mileage warranty limits
- Effect of the pattern of failure rates and mileage accumulation rates on the warranty cost

OVERVIEW

This chapter provides a simple method to assess the impact of new time/mileage warranty limits on the number and cost of warranty claims for components and subsystems of a new product. Section 8.1 discusses the changes in the warranty coverage that are occurring in the marketplace. Section 8.2 provides a method to arrive at the number of claims per 1000 vehicles in the field for new time/mileage warranty limit combinations and illustrates the method with an example. It also highlights the use of mileage accumulation rates of a population of vehicles to arrive at claims per 1000 vehicles, sold with new time/mileage warranty limits. Section 8.3 discusses the bias in warranty cost estimates that may result in using cumulative cost per repair information, and recommends the use of incremental cost per repair, especially when populations with different mileage accumulation rates are under consideration. Application examples illustrate the use of the methodology.

8.1 CHANGES IN THE WARRANTY COVERAGE

Two-dimensional warranty coverage is generally stated in terms of mileage and time (in months or years) limits. It expires when any of the two limits is crossed. Warranty

---

limit denoted as M_2/M_1 would henceforth refer to the base warranty coverage meaning M_2 months or M_1 usage (in 1K mi), whichever occurs first. Any new warranty limit of interest is denoted as $M_2^{'}/M_1^{'}$.

During the last decade, warranty coverage for an automobile has changed from 12/12, that is, 12 months or 12K mi, to 36/36, that is, 36 months or 36K mi. However, many automakers in an attempt to improve their competitive sales position are beginning to offer more than 50/50 coverage. As a result of such changes, a company needs to proactively plan for maintaining a large cash reserve to pay for the warranty services on their products. Kakouros et al. (2003) point toward the risk of adopting a presumably more attractive warranty policy in order to compete in the market, without being able to assess the cost associated with the newly adopted policy. Figure 8.1 shows different scenarios that need to be assessed before choosing a new time or mileage limit.

This chapter discusses one such quantitative methodology to assess the impact of changing M_2/M_1 to any new coverage $M_2^{'}/M_1^{'}$ for a new vehicle (Rai and Singh 2005a). Data from the previous model-year vehicle that are available through an automotive warranty database is used for the assessment.

8.2   THE ESTIMATION OF NUMBER OF WARRANTY CLAIMS

The warranty cost for any coverage, M_2/M_1, depends on the number of claims and cost of repairing each claim. In this section, we develop a method of estimating the number of claims for a specific subsystem of a vehicle at various combinations of time/mileage limits of warranty. Although the method focuses on analysis at the component/subsystem level, the estimates at the system or vehicle level can be easily obtained by the pooling of component- or subsystem-level estimates. The main objective of the estimation method being discussed is to be able to assess the impact of changes in time and/or mileage limits on the number of claims. For this purpose, in the following, we

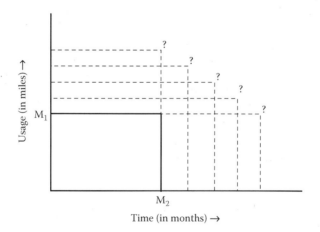

FIGURE 8.1   New time/mileage warranty limits.

discuss the issues of choice of data set, modeling of first failures, modeling of mileage accumulation, repeat failures, and claims per 1000 vehicles sold.

8.2.1   THE CHOICE OF A DATA SET

Choosing an appropriate data set for analysis is the first step toward making an assessment of the impact of changes in time/mileage limits. The choice depends upon whether the subsystem under study will be a part of the newly designed vehicle or a new model of an existing vehicle. However, it is important to identify a representative data set from those available for the analysis. Choosing a data set belonging to a current model-year vehicle may lead to complex truncation and censoring mechanisms in the data set that pose great difficulty in arriving at estimates for even the base warranty period. Low sample size for vehicles with higher time-in-service is also a hindering factor. Moreover, there are chances that failure modes expected at the later part of the life cycle of a vehicle may not get any representation at all. This could lead to inefficient estimates for the expected number of claims at any given month-in-service (MIS) value.

Similarly, choosing a data set of a very old model-year vehicle could involve outdated failure modes due to design improvements over various model-year vehicles. Although sample size may not be a problem, such data may not constitute a representative data set. Choosing a model-year vehicle known to have an unusually high number of special causes due to design or manufacturing issues and resulting in inflated figures of warranty claims should also be avoided. Such a data set will give rise to highly unstable estimates with poor statistical properties. A reasonable choice could be a recent model-year vehicle where the majority of the vehicles in the population completed a substantial part of the base warranty period and a model-year vehicle that did not have significantly high number of special cause-related issues.

8.2.2   MODELING OF THE FIRST FAILURES

Once a representative data set is chosen, the next step is to model the first failures. Figure 8.2 shows a mapping of occurrence of failures (in time in months and usage in miles) for four cars.

Figure 8.2 shows that the failures occurring at points 1, 2, 3, and 6 generally result in a warranty claim and are thus captured by the warranty data. The failures occurring at points 1 and 6 represent the first warranty claims for the respective vehicles. The repeat claims represented by points 2 and 3 will be discussed after addressing modeling and analysis of the first failures. The failures occurring outside the warranty period, such as those at points 4, 5, 7, and 8, are not captured by the warranty data set, but need to be estimated based on the data available within the warranty limits.

In the preceding chapters, several methods for arriving at nonparametric hazard rate estimates for hard and soft failures were discussed. However, assessing the impact of a new time/mileage warranty limit requires hazard rate projections beyond the time or mileage for which the data are available. This is done by modeling the hazard rate estimates using statistical distributions such as Weibull, lognormal,

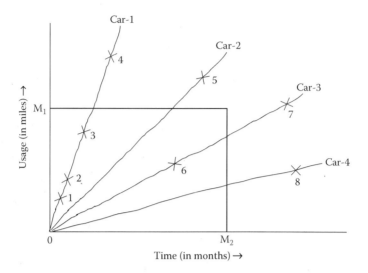

FIGURE 8.2   Occurrence of failure in a subsystem of a car.

exponential, etc., and then using the parameters to go beyond the existing time or mileage. Modeling of mileage accumulation rates in the vehicle population that is required for making such projections is discussed in Chapter 4, Section 4.2. Nelson (1982) in Chapter 4 provides a hazard-plotting method to obtain distribution parameters of the failure density function.

8.2.3   REPEAT FAILURES

For any given subsystem, it is expected that as MIS for a population of a specific type of car increase, the number of repeat claims as a proportion of first claims would also increase. In other words, it can be regarded as a nondeceasing function of time. Repeat claims could be the result of either a new failure or difficulty in root cause elimination during the previous repair. For example, a car that had a new engine installed because of engine failure may again experience the same problem owing to manufacturing- or assembly-related issues. On the other hand, a car previously repaired for excessive engine noise may again return for the same repair owing to inaccurate diagnosis during the previous visit.

The expected number of total claims can be obtained by combining estimates of repeat claims with the estimates for the first claims. Repeat claims as a proportion of the first claims can be estimated using the following formula:

$$\hat{p}_{t,}M_1 = \frac{n_T(t) - n_f(t)}{n_f(t)} \tag{8.1}$$

where

$\hat{P}_{t,M_1}$: Estimate of repeat claims as a proportion of the first claims at MIS = t with M_1 as the mileage warranty limit.

$n_T(t)$: Total number of claims up to MIS = t.

$n_f(t)$: Number of first claims up to MIS = t or number of cars with at least one claim up to MIS = t.

EXAMPLE 8.1

Consider the number of claims per vehicle, number of vehicles, and total number of claims at $t = 15$ as given in Table 8.1.

As there are 100 vehicles with 130 claims, $n_f(15) = 100$ and $n_T(15) = 130$. Therefore,

$$\hat{P}_{15,M_1} = \frac{130 - 100}{100} = 0.30$$

Thus, there are 30% repeat claims as a percentage of first claims at MIS = 15.

After obtaining $\hat{P}_{t,M_1}$ values at $t = 1, 2, ..., M_2$ within the warranty coverage region, a curve is fitted to capture the nondecreasing trend for the repeat claims. When increment of 5 or 10 MIS is used for arriving at $\hat{P}_{t,M_1}$ values, the curve fitted to the data points can be used to arrive at intermediate $\hat{P}_{t,M_1}$ values. Using $\hat{P}_{t,M_1}$ and parameters of $G(w_t)$, $\hat{P}_{t,unlimited}$ is estimated as:

$$\hat{P}_{t,unlimited} = \frac{\hat{P}_{t,M_1}}{P[w_t < M_1]} \tag{8.2}$$

where $\hat{P}_{t,unlimited} \geq \hat{P}_{t,M_1}$.

A curve fitted to $\hat{P}_{t,unlimited}$ versus t is used in conjunction with $G(w_t)$ to estimate $\hat{P}_{t,M_1}$ for any new mileage limit M_1. In the absence of any data on repeat claims beyond the existing base warranty period, it is assumed that the repeat claims would continue to grow at a rate established by the fitted polynomial. An example of $\hat{P}_{t,36}$ and $\hat{P}_{t,unlimited}$

TABLE 8.1
Example of Repeat Claims at MIS = 15

Claims per Vehicle	Number of Vehicles	Number of Claims
1	80	80
2	10	20
3	10	30
Total	100	130

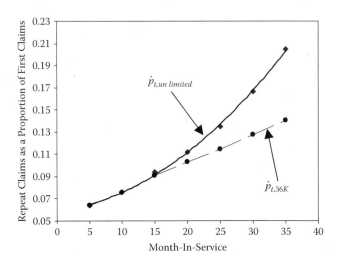

FIGURE 8.3   Repeat claims as a proportion of first claims.

for a subsystem of a car is shown in Figure 8.3. It shows that $\hat{p}_{t,un\,limited} \geq \hat{p}_{t,M_1}$ for all t. It also shows that up to about 15 MIS both values are very close as during this period the probability of a car going out of a 36K-mi warranty is almost negligible. It is also to be noted that a curve for any new mileage limit $M_1 > 36K$ mi for a population of cars with the same mileage accumulation rates, would lie in between the two curves.

8.2.4   Claims per Thousand

Using $f_T(t)$ and $G(w_t)$, the number of the first warranty claims per 1000 cars sold with a new warranty coverage M_2 / M_1 can be obtained at each MIS $= t$ as

$$\hat{m}_{first}(t, M_1) = f_T(t) \times P[W_t < M_1] \times 1000 \qquad (8.3)$$

where $f_T(t) = F(t) - F(t-1)$

   Cumulative number of first claims for every 1000 cars sold with warranty coverage M_2 / M_1 can be obtained at each MIS $= t$ as

$$\hat{M}_{first}(t, M_1) = \sum_t m_{first}(t, M_1) \qquad (8.4)$$

Finally, the expected number of total claims per 1000 vehicles sold (C/1000) with a new warranty coverage M_2 / M_1 is obtained using,

$$\hat{M}_{Total}(M_2^{'},M_1^{'})= \hat{M}_{First}(M_2^{'},M_1^{'})+\left\{\hat{M}_{First}(M_2^{'},M_1^{'})\times \hat{p}_{M_2^{'},M_1^{'}}\right\} \qquad (8.5)$$

where $\hat{p}_{M_2^{'},M_1^{'}} =\left(p_{M_2^{'},un\lim ited}\right)\times P\left(W_{M_2^{'}}<M_1^{'}\right)$

C/1000 is analogous to repairs per 1000 or R/1000 terminology used in industries. Very often for a single claim there are multiple repairs done on the same vehicle. R/1000 does not capture all the multiple repairs; in reality, it captures only the number of claims made per 1000 vehicles sold. Thus, the term "repairs/1000" is a misnomer and "claims/1000" is a more appropriate term to use.

8.2.5 APPLICATION EXAMPLE 1

The expected numbers of claims for different combinations of time and mileage limits were to be assessed for a specific subsystem of a new engine design, say "X." The design is similar to a previous model-year vehicle, say "A," and is similar in application to another model-year vehicle, say "B," on which it will be used. Due to design similarity, nature of failure modes for the newly designed subsystem is expected to be similar to "A." Accordingly, for modeling the first and the repeat claims, the data set involving "A" is chosen. However, due to the application similarity to "B," mileage accumulation data available for vehicle B is used.

To enable a clearer understanding, consider the data on warranty claims to be the failure of a sealing material to prevent transmission fluid leaks in a vehicle. For the subsystem of vehicle A containing the sealing material, Weibull distribution (parameters: shape = 0.95, location = 1840) provided a good fit to the first warranty claims (Murthy et al. 2004). The distribution parameters are obtained using the hazard-plotting method discussed in Nelson (1982). Mileage accumulation rates for vehicle A has the best-fit distribution as lognormal (parameters: location = 6.6646, scale = 0.625), with mean and sigma of 953 mi/month and 659 mi/month, respectively. Similarly, mileage accumulation rates for vehicle B has the best fit distribution as lognormal (parameters: location = 7.1, scale = 0.45) with mean and sigma of 1341 mi/month and 636 mi/month, respectively. It is to be noted that vehicle B has a higher mileage accumulation rate as compared to vehicle A.

Using pdf $g_A(w_t)$ and Equation 8.1, an estimate of $\hat{P}_{t,un\lim ited}$ is arrived at and is shown in Figure 8.3. A second-degree polynomial fitted with $R^2 = 0.9988$ is as follows:

$$\hat{P}_{t,un\lim ited} = 0.0001\ t^2 + 0.0005\ t + 0.0596 \qquad (8.6)$$

An estimate of number of first claims per 1000 cars sold for different combinations of warranty coverage $M_2^{'}/M_1^{'}$ is obtained from Equation 8.3 using $g_B(w_t)$. Subsequently, $\hat{M}_{first}(M_2^{'},M_1^{'})$ is obtained for each $M_2^{'}/M_1^{'}$ combination. $\hat{M}_{Total}(M_2^{'},M_1^{'})$ is then obtained for each $M_2^{'}/M_1^{'}$ combination from Equations 8.5 and 8.6. Comparison of actual $M_{Total}(36,36)$ and estimate $\hat{M}_{Total}(36,36)$ for vehicle A showed an error of 9.3%, which is regarded as reasonable enough to proceed with the assessment.

Although the subsystem under study for the new vehicle X and previous model-year vehicle A are similar in design, the absolute values of $\hat{M}_{Total}(M_2', M_1')$ for the two vehicles are not expected to be the same. One of the main reasons is the improvement in the design of sealing material carried out for vehicle X over a period of time. This implies that the results obtained should be used for assessing the effect of different time/mileage warranty limit combinations and not for arriving at an absolute value of the estimates concerning vehicle X.

For making the assessment, assume that the design target for the subsystem under study for vehicle X is $M_{Total}(36,36)=10$. Using the design target information and the results obtained so far, $\hat{M}_{Total}(M_2'/M_1')$ in terms of design targets for different time/mileage warranty limits are obtained using

$$\frac{\hat{M}_{Total}(36,36)}{\hat{M}_{Total}(M_2'/M_1')} \times Design\ Target \qquad (8.7)$$

The estimates for C/1000 using a design target of 10 are given in Figure 8.4.

It is observed from Figure 8.4 that the rate of increase in claims is higher when mileage limit is increased for a given time limit. When the time limit is increased for a given mileage limit, the rate of increase in C/1000 is lower. One of the main reasons for such a trend is the high mileage accumulation rate (over 1300 mi/month) for vehicle B. Impact of mileage accumulation and warranty time/mileage limits on C/1000 can also be seen from Figure 8.5.

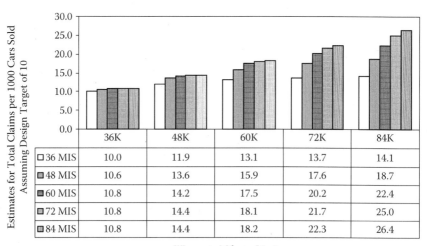

	36K	48K	60K	72K	84K
□ 36 MIS	10.0	11.9	13.1	13.7	14.1
▦ 48 MIS	10.6	13.6	15.9	17.6	18.7
▨ 60 MIS	10.8	14.2	17.5	20.2	22.4
□ 72 MIS	10.8	14.4	18.1	21.7	25.0
▦ 84 MIS	10.8	14.4	18.2	22.3	26.4

Warranty Mileage Limit

FIGURE 8.4   Estimated C/1000 for X-type car at different time/mileage combinations.

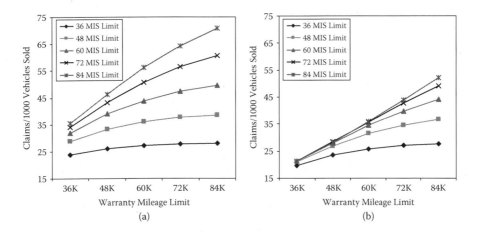

FIGURE 8.5   Effect of mileage accumulation rates on C/1000: (a) Vehicle A (950 mi/month), (b) Vehicle B (1340 mi/month).

It is observed from Figure 8.5a that the curve corresponding to the 36 MIS limit is almost flat across mileage limits. This is due to the fact that majority of cars of type A would go out of warranty coverage due to crossing the warranty time limit rather than the mileage limit. Note that the mileage accumulation rate of about 950 mi/month for vehicle A is similar to car 3 in Figure 8.2, which crosses time limit before the mileage limit. For vehicle A, at 36 MIS, 65% of the population of cars would still have mileage less than 36K mi, and about 95% mileage less than 84K mi. Consequently, as the warranty time limit is raised, the rate of increase observed in C/1000 becomes higher across mileage limits. In other words, increase in time limit impacts C/1000 more than the increase in warranty mileage limit.

It is observed from Figure 8.5b that the estimates for C/1000 at the 36K mileage limit are very close to each other for warranty time limits from 36 to 84 months. The main reason is that more cars in the population are expected to be out of the warranty limit due to crossing of mileage limit rather than the time limit. Again, note that mileage accumulation of about 1340 mi/month for vehicle B is similar to that of car 2 in Figure 8.2, which crosses the mileage limit before time limit. For vehicle B, at the 36K mileage limit, only 33%, 14%, 6%, 2.5%, and 1% of the population of cars will still be within the time limit of 36, 48, 60, 72, and 84, respectively. Thus, raising the warranty time limit does not have a significant impact on the C/1000 at lower values of the mileage limit. However, as the mileage limit is increased, a more significant impact on C/1000 is observed for the increasing warranty time limit.

The comparison in Figure 8.5a,b shows that for higher mileage accumulation rates, the overall C/1000 is lower than those expected for lower mileage accumulation rates. However, increase in expected C/1000 is steeper across increasing mileage limits for the higher mileage accumulation rates than that for the lower. Thus, mileage accumulation rates for the population of vehicles play a key role in assessing effect of changing time/mileage limits on C/1000.

In the next section, a method for estimating warranty cost per vehicle sold for a given subsystem is discussed.

8.3   COST OF WARRANTY CLAIMS

The cost per unit sold mainly includes time taken by the service technician to successfully repair the vehicle and the material cost. The absolute value of cost per repair may also depend on the nature of failure mode. For example, the cost of repairing an oil level indicator assembly may not exceed $25. On the other hand the cost to repair a vehicle experiencing an engine failure may run into a few thousand dollars.

Obtaining warranty cost per unit sold for a subsystem requires estimates of expected C/1000 and the cost per repair at different MIS values. A method for estimating the former was discussed in previous section. In this section, a method for arriving at warranty cost per unit sold using estimates for C/1000 and incremental cost per repair (ICPR) is discussed.

Incremental cost per repair $c_r(t)$ at any MIS $= t$ ($t = 1, 2 \dots , M_2$) obtained from base warranty period M_2/M_1 is given by,

$$c_r(t) = \frac{C(t) - C(t-1)}{n_T(t) - n_T(t-1)} \tag{8.8}$$

where $C(t)$ is total warranty cost up to t, $n_T(t)$ is total warranty claims up to t, and $C(0) = n_T(0) = 0$.

A curve fitted to $c_r(t)$ versus t captures the changes in warranty cost per repair as a function of MIS. There are numerous factors that impact cost of repairs. Some of the major factors are nature and time of the occurrence of a failure mode, mileage accumulation rates for a population of cars, and the design actions such as design for serviceability. The first two factors are discussed in further detail along with a few examples.

8.3.1   THE EFFECT OF FAILURE NATURE AND TIME OF OCCURRENCE ON WARRANTY COST

The parametric model fitted to the data gives information regarding nature of the failure mode under study. For example, when Weibull distribution provides a good fit to the first warranty claims in the data set, its shape (b) and location (q) parameters give useful information about the nature of the failure mode for the subsystem under study. Decreasing, constant, and increasing failure rate phases of a subsystem life cycle is captured using Weibull distribution by $b < 1$, $b = 1$, and $b > 1$, respectively. For example, during transmission assembly, an improper alignment of the sealing material may damage the seal, causing fluid leaks. If such manufacturing- or assembly-related defects reach customers, decreasing failure rate with $b < 1$ is more likely to occur where majority of such claims are made during the initial life cycle of the vehicle population. On the other hand, excessive usage of the vehicle may lead to wearing out or degradation of the sealing material, resulting in transmission fluid leaks. Such a phenomena is more likely to reflect the $b > 1$ situation. Similarly, cost per repair for the same failure mode may be different for the $b < 1$ situation as compared to the $b > 1$ situation.

The effect of the parameters b and q on cumulative cost per repair (CCPR) for a subsystem is shown in Figures 8.6 and 8.7, respectively, where CCPR denoted by $CC(t)$ is given by,

$$CC(t) = \frac{\sum_t \left[\hat{M}_{Total}\left(t, M_1'\right) - \hat{M}_{Total}\left(t-1, M_1'\right) \right] \times c_r(t)}{\hat{M}_{Total}\left(t, M_1'\right)} \tag{8.9}$$

Figure 8.6 shows that CCPR changes when b is changed from 0.5 to 1.5. For $b = 0.5$, which relates to decreasing failure rate situation, CCPR is higher from about 5 to 25 MIS compared to the other two curves. This is due to the higher contribution of the number of failures at the beginning of the vehicle life cycle. Whereas for $b = 1.5$, which relates to increasing failure rate situation, CCPR is lower in initial months but increases significantly at the higher MIS as the contribution of the number of failures increases. For $b = 0.95$, which is closer to constant failure rate situation, the CCPR remains in between the two extremes.

Figure 8.7 shows the effect of changing q while holding b constant at 0.95. For lower value of q, CCPR is slightly higher in the initial period of MIS. This is due to the fact that majority of failures occur early in this case. Thus, CCPR for $q = 18.4$ remains almost constant after 30 MIS. Whereas for $q = 184$ and 1840, when majority of failures occur at later period, CCPR shows an increasing trend.

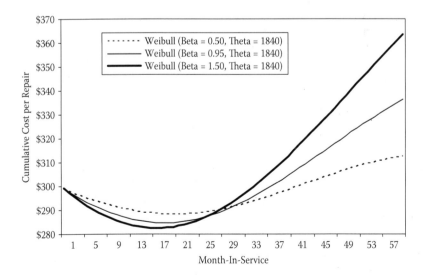

FIGURE 8.6   Effect of b on CCPR with mileage accumulation of 1340 mi/month, and $M^1 = 60$.

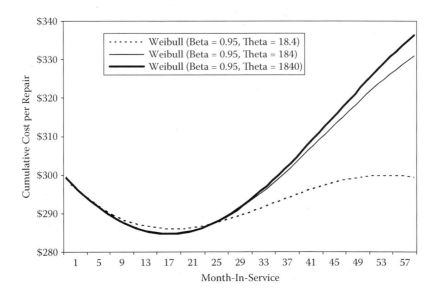

FIGURE 8.7   Effect of q on CCPR with mileage accumulation of 1340 mi/month, and $M^1 = 60$.

Apart from the nature of a failure mode, the time or vehicle age at which a failure occurs also influences the cost of repair. For example, engine oil leak experienced in a new car very often may require only a minor repair to overcome the problem. However, to repair the same problem in an older car, which may be a result of multiple engine oil leaks, the engine itself may require replacement. Such factors can lead to higher costs of repair in older vehicles.

8.3.2   The Effect of Mileage Accumulation Rate on Warranty Cost

A population of cars with relatively high mileage accumulation rates may not experience time-related failure modes within the warranty period owing to crossing of the mileage limit early. Similarly, a population of cars with relatively lower mileage accumulation rates may not experience some of the high usage-related failure modes. This can lead to potentially different trends for cost of each repair over a period of time. The differences in the mileage accumulation rates may affect warranty cost, even though the design may be similar. An example is shown in Figure 8.8.

It is observed from Figure 8.8 that up to about MIS = 35, CCPR are very close to each other for different mileage accumulation rates. However, beyond MIS = 35 differences in CCPR become more prominent. High CCPR is obtained when mileage accumulation is minimum and vice versa. Lower CCPR for higher mileage accumulation rates is mainly due to less contribution of vehicles that failed beyond 35 MIS. Thus, CCPR based on the base warranty and original mileage accumulation rate shows an underestimation of the CCPR for a lower mileage accumulation rate of 545 mi/month. Similarly, at the higher mileage accumulation rate of 1341 mi/month,

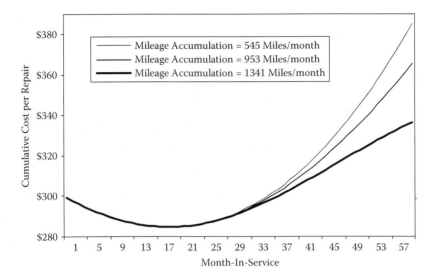

FIGURE 8.8 Effect of mileage accumulation on CCPR when first failures follow Weibull distribution with b = 0.95 and q = 1840 with = 60.

CCPR based on base warranty and original mileage accumulation rate shows an overestimate of the cost.

These results indicate that CCPR obtained from warranty period of M_2/M_1 (which in this case is 36/36) should not be used directly in the calculation of warranty cost per unit sold. The results will not be significantly biased if the new design being assessed has similar mileage accumulation and failure rates to the representative model-year vehicle chosen. However, the results could be highly biased if the mileage accumulation and failure rates differ significantly. Thus, to avoid bias in the estimation of warranty cost, ICPR as given in Equation 8.8 should be used for arriving at warranty cost per unit sold.

Warranty cost per unit at any MIS = t is obtained using estimates of C/1000 and incremental cost per repair as follows,

$$WC(t) = \frac{\sum_t \left[\hat{M}_{Total}\left(t, M_1'\right) - \hat{M}_{Total}\left(t-1, M_1'\right) \right] \times c_r(t)}{1000} \tag{8.10}$$

The next section extends the example discussed in application example 1 for arriving at warranty cost per unit sold for new time/mileage warranty limit combinations.

8.3.3 APPLICATION EXAMPLE 2

In application example 1, the estimates for $\hat{M}_{Total}(M_2', M_1')$ were obtained. This example makes use of these estimates to assess effect of changing time/mileage warranty limits on warranty cost per unit sold. From the data set for vehicle A with warranty coverage of 36/36, ICPR is obtained for $t = 1, 2, \ldots, 36$ using Equation 8.8. A second-degree polynomial fitted with $R^2 = 0.8971$ is as follows:

$$c_r(t) = 0.1474t^2 - 3.7492t + 302.84 \qquad (8.11)$$

In the absence of data on ICPR beyond 36/36, it is assumed that it will continue to increase at a rate captured by Equation 8.11. The increasing trend in ICPR obtained from Equation 8.11 indicates that at a higher time/mileage, failures due to wearing out would cause cost of each repair to be higher.

Using Equations 8.10 and 8.11, and mileage accumulation for vehicle B, $WC(t)$ for vehicle X is obtained. Assuming a design target of $1 at M_2/M_1 for the subsystem of vehicle X under study, estimates for total warranty cost per car sold are obtained using

$$\frac{\hat{WC}(36,36)}{\hat{WC}(M_2',M_1')} \times Design\ Target\ (= \$1) \qquad (8.12)$$

The estimates for warranty cost per car sold using a design target = $1 for the subsystem is given in Figure 8.9.

It is observed from Figure 8.9 that, as compared to the base warranty period of 36/36, the estimated warranty cost per car sold increases to 1.4 times for the 36/84 warranty coverage. On the other hand, the increase is about four times when warranty coverage changes from 84/36 to 84/84. It is also observed that from 36/36 to

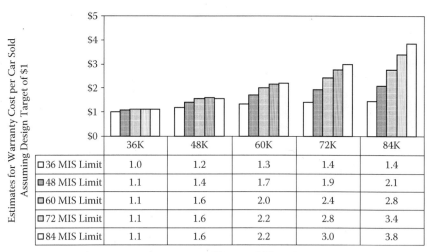

	36K	48K	60K	72K	84K
☐ 36 MIS Limit	1.0	1.2	1.3	1.4	1.4
■ 48 MIS Limit	1.1	1.4	1.7	1.9	2.1
☐ 60 MIS Limit	1.1	1.6	2.0	2.4	2.8
☐ 72 MIS Limit	1.1	1.6	2.2	2.8	3.4
☐ 84 MIS Limit	1.1	1.6	2.2	3.0	3.8

Warranty Mileage Limit

FIGURE 8.9 Estimated warranty cost per unit of type X for different time/mileage warranty limit combinations.

84/36, the increase in warranty cost per unit sold is almost negligible, and from 36/84 to 84/84, the increase in warranty cost per unit is slightly under three times. This shows that the increasing warranty mileage limit for vehicle X will increase the warranty cost per unit more compared to increase in the warranty time limit. Once such an assessment is made for all the subsystems of the vehicle; by pooling the estimates, an overall assessment of the vehicle X for different time/mileage limits can be made.

8.4   SUMMARY

Key points from this chapter are summarized as follows:

- This chapter provided a simple method to assess the impact of new time and mileage limits. The method discussed in this chapter will be found handy when quick estimates for the number and/or cost of warranty claims for a new coverage are required.
- It was demonstrated that the mileage accumulation rates in the population of vehicles have a significant impact on the number of claims. The rate of increase in the number of claims is different over changing warranty time/mileage limits when mileage accumulation rates are different. In general, when mileage accumulation rates are high, the expected number of claims is lower than that for low mileage accumulation rates. It is, however, observed that the rate of increase in the number of claims is steeper for an increasing mileage limit when the mileage accumulation rates in the population is higher.
- The effect of mileage accumulation and failure distribution parameters on cumulative cost per repair is also discussed with examples. It is found that when incremental cost per repair is used to arrive at the cumulative cost per repair, different cumulative costs per repair (CCPR) values are obtained for different mileage accumulation rates of a population of cars. This indicated that using a fixed CCPR based on base warranty period might result in biased estimates. An example with fixed mileage limit and varying mileage accumulation rates showed that CCPR based on base warranty for a given mileage accumulation will lead to an underestimate when actual mileage accumulation rates are lower and vice versa. Thus, the use of incremental cost per repair is more appropriate than cumulative cost per repair, especially when populations of vehicles with different mileage accumulation rates are under consideration.
- The method discussed in this chapter should be used for comparative studies only. The absolute values of estimates cannot be used for predicting number or cost of claims at a future date. This is because the data set for analysis is selected from a previous model-year vehicle, so as to minimize complex censoring mechanisms present especially in a latest model-year vehicle. Thus, the design improvements that take place over the years may not get captured in the chosen data set. Another issue is the possibility of

new failure modes at a later part of the vehicle life cycle, which also are not captured by warranty data set.

- For the new model-year vehicle, the quantitative effect of design improvement actions is not known until sufficient warranty data gets accumulated. However, the design engineers do have qualitative information about their actions for the design improvements. To incorporate such information in the estimation process, the use of fuzzy logic and Bayesian methods can be helpful (Yadav et al. 2003). However, further work is required to extend the methodology discussed in this chapter to include the effect of design improvement actions.

BIBLIOGRAPHIC NOTES

Nonparametric and parametric methods that help to study repeat claims data (or recurrence data in general) are very often useful for studying warranty claims/costs. Nelson (2003) provides a unified theory for arriving at nonparametric estimates involving recurrence data that apply not only to "counts" but includes "costs" and other "values" of discrete events. Lawless (1995) and Lawless and Nadeau (1995) also provide methods for analysis of recurrent events. Karim et al. (2001) present a nonhomogeneous Poisson process (NHPP) model for repairable products and a multinomial model and its Poisson approximation for nonrepairable products. Lawless (1987) provides regression methods for Poisson process data.

9 Forecasting of Automobile Warranty Performance

Prediction is very difficult, especially about the future.

—**Niels Bohr**

OBJECTIVES

This chapter covers the following:

- The need for warranty performance forecasting
- The phenomena of "warranty growth"
- Existing methods and their limitations
- Warranty forecasting using neural networks
- A method for optimizing network parameters
- Comparison of forecasting performance using log–log plot, radial basis function (RBF), and multilayer perceptron (MLP) networks

OVERVIEW

This chapter describes the application of neural networks to forecast warranty performance in the presence of "warranty growth" phenomena. Section 9.1 introduces the phenomena of warranty growth. Section 9.2 discusses warranty performance forecasting using log–log plots and dynamic linear models. Sections 9.3 and 9.4, respectively, provide RBF and MLP neural network methods for forecasting the warranty performance. To optimize the network parameters, training and testing errors are minimized through planned experimentation. Section 9.5 discusses the results obtained. The key points are summarized at the end of the chapter in Section 9.6.

9.1 WARRANTY GROWTH OR MATURING DATA PHENOMENA

An automobile with over 7000 parts is a highly complex product. In spite of employing the best quality and reliability practices during product development, manufacturing, and assembly, unexpected failures during the warranty period do occur and cost automobile companies billions of dollars annually in warranties alone. Warranty cost reduction programs in these companies thus receive high priority. Various teams work together to achieve objectively defined targets for warranty cost reduction that

are often based on the performance of the previous model-year vehicles. Two important measures of automobile warranty performance are

1. The cumulative number of repairs carried out per 1000 vehicles in the field.
   It is usually denoted by R/1000 and is calculated as

$$R/1000 = \frac{\text{Total number of repairs up to time } t}{\text{Total number of vehicles in the field at time } t} \times 1000$$

2. The cumulative cost of repair per vehicle in the field. This measure is
   denoted by CPU, meaning cost per unit, and is calculated as

$$CPU = \frac{\text{Total warranty cost up to time } t}{\text{Total number of vehicles in the field at time } t}$$

The year-end R/1000 and CPU targets are generally defined for both low months in
service (LMIS) and high months in service (HMIS). The LMIS of interest could be
target R/1000 and CPU values at 1, 3, 6 months in service (MIS) for vehicles in the
field. Similarly, HMIS of interest could be target R/1000 and CPU values at 12 MIS
or any other higher MIS of interest. Each month, as new vehicles are sold, the vehicle
population increases. Similarly, with each additional MIS spent by the vehicle population, cumulative number of repairs, too, increases. In order to proactively capture
the gap between target and actual performance, the engineers forecast R/1000 and
CPU numbers that would be achieved by the end of the calendar year.

Predictions that are significantly higher than the actual repairs/complaints may
result in unnecessarily costly design, manufacturing, or service actions. On the other
hand, predictions that are significantly lower than the actual repairs/complaints may
lead to false confidence of meeting the year-end targets for warranty performance,
thus resulting in higher than expected warranty costs. When forecasting is done at
the end of the first quarter of a calendar year, it involves a 9-months-ahead prediction. On the other hand, forecasting at the beginning of the last quarter involves
up to a 3-months-ahead prediction. The scope of such forecasts varies from major
assembly such as transmission or engine to subassembly such as turbocharger, torque
converter, or pump. Occasionally, the focus is also directly on a major customer
concern such as fluid leaks, engine noise, lack of power, etc. Apart from helping
the automobile engineers to fine-tune their strategies for warranty cost reduction
through design, manufacturing, or service actions, a sound forecasting method also
helps the company to plan for the unexpended warranty cost to be paid.

In this chapter, we study the forecasting of warranty performance using R/1000
values. Forecasting of CPU values can be carried out in a similar fashion. An example of R/1000 for a given model-year vehicle after 9 MIS is shown in Figure 9.1.

It can be seen from Figure 9.1 that the R/1000 values increase in magnitude as
MIS values increase. This is due to the cumulative nature of the warranty performance measure. To forecast R/1000 values beyond 9 MIS, one of the simplest methods used is log–log plot method. Here, R/1000 and MIS values are simply plotted on

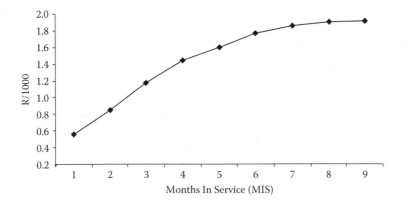

FIGURE 9.1   R/1000 values for a given model-year vehicle after 9 months in service.

logarithmic scales where they look approximately linear. An appropriate curve fitted to such a plot is then extrapolated to predict R/1000 values at a future MIS value of interest. Chen et al. (1996) used similar data from several previous model-year vehicles along with the current model-year data to forecast R/1000 values at future MIS using dynamic linear models. However, the R/1000 values for a given MIS are known to change when higher MIS values become available. This is depicted in Figure 9.2. Note that the curve shown in Figure 9.1 is the curve representing month 9 in Figure 9.2.

Figure 9.2 shows R/1000 values up to 9 months since the beginning of the vehicle model year with MIS values from one to nine. It can be seen that MIS-1 R/1000 value at month 2 is higher than it was at month 1. Similarly, MIS-1 R/1000 at month

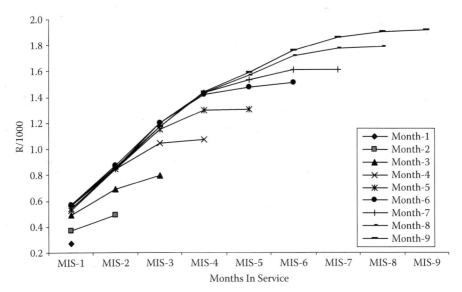

FIGURE 9.2   R/1000 values over various months in service.

3 is higher in comparison to its value at month 2. Such "maturing" of R/1000 values at MIS-1 continues with reduced magnitude of R/1000 increase, as the data from more months become available. The same is true for the remaining R/1000 values at each MIS.

It is important to capture such maturing of R/1000 values in order to accurately forecast the R/1000 values at future months. Several authors, including Robinson and McDonald (1990) and Singpurwalla and Wilson (1993), point out this phenomenon of maturing data or warranty growth. It is easy to visualize that the forecasts obtained using log–log plot method are likely to underestimate the true R/1000 value. Singpurwalla and Wilson (1993) note that due to the data maturation phenomena, graphical methods of extrapolation based on log–log plot are particularly ill suited. They instead suggest that dynamic linear models with innovation terms that account for the added uncertainties due to a maturation of the data to be a better approach.

While it is possible to develop statistical methods that address the issue of maturing data to forecast the year-end performance, we employ neural networks for certain advantages they possess. Some of the important advantages of using neural networks for forecasting problems are as follows:

- They can be trained to find complex relationships in data (Wedding and Cios 1996).
- They allow modeling of nonlinear data using a linear approach (Yao et al. 2002; Vojinovic et al. 2001).
- Networks such as RBF neural networks have a compact and simple network structure with a small number of neurons (Shi et al. 1999; Korres et al. 2002; Xu et al. 2003).
- Networks such as RBF neural networks have small training times (Shi et al. 1999; Korres et al. 2002; Yao et al. 2002; Xu et al. 2003).
- Networks such as MLP neural networks use supervised learning using a highly popular back-propagation algorithm (Haykin 1994).

9.2 CURRENT WARRANTY FORECASTING METHODS[*]

There are a large number of variables that influence the patterns/trends observed in warranty performance, and it is practically infeasible to develop a model that includes all the variables. Some of the variables that relate to the development of a warranty-forecasting model are captured in Figure 9.3.

In Figure 9.3, control factors indicate those variables over which the model builder can exercise full control. The forecasting method chosen by the model builder has a large influence on the forecasting accuracy. The decision for selecting a particular forecasting method may take into account constraints such as time, accuracy needed, model complexity, etc. The application of any forecasting method becomes difficult

[*] © 2005 Taylor & Francis Group Ltd. This section is reprinted with permission from: Rai, B. K. and Singh, N. (2005). Forecasting warranty performance in presence of maturing data phenomena. *International Journal of System Sciences*, 36(7), 381–394..

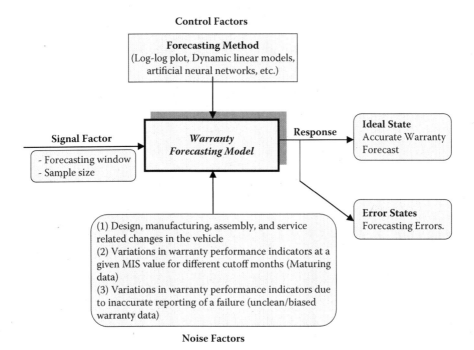

Control Factors

Forecasting Method
(Log-log plot, Dynamic linear models,
artificial neural networks, etc.)

Signal Factor
- Forecasting window
- Sample size

Warranty Forecasting Model

Response

Ideal State
Accurate Warranty Forecast

Error States
Forecasting Errors.

(1) Design, manufacturing, assembly, and service related changes in the vehicle
(2) Variations in warranty performance indicators at a given MIS value for different cutoff months (Maturing data)
(3) Variations in warranty performance indicators due to inaccurate reporting of a failure (unclean/biased warranty data)

Noise Factors

FIGURE 9.3   Factors influencing warranty forecasting model.

due to the presence of noise factors, over which the model builder does not have much control. Noise factors are mainly responsible for the error state that, in this case, translates to forecasting errors. In an ideal situation, if there is no variation due to noise factors, forecasting accuracy will be almost perfect.

Signal factors are those that have a large influence on the response. Forecasting errors for the immediate future—say, month 15—are likely to be lower when compared to forecasting errors for the distant future—say, month 30—when data up to month 14 is available. Similarly, forecasting errors with a very small data set (e.g., when data is available only up to month 5) is likely to be larger as compared to forecasting errors with a larger data set (when the data is available up to month 15).

We now briefly discuss certain advantages and shortcomings of two warranty forecasting methods used or referenced in the literature.

Log–log plots. This is one of the simplest methods to predict warranty performance. It is achieved by simply plotting a warranty performance indicator such as R/1000 and MIS values on logarithmic scales where they look approximately linear, and then extrapolating. For example, if we have data up to month 14, then the maximum available value for MIS is 14 months. The R/1000 values for MIS = 1 through MIS = 14 for month 14 can be plotted on a log–log scale. An appropriate curve fitted to such a plot can then be extrapolated to predict R/1000 values for MIS = 15 expected at month = 15 and MIS = 16 expected at month = 16. However, it is easy to visualize that the forecasts obtained in this fashion are likely to underestimate the true R/1000 value as this method ignores the phenomena of maturing data. Singpurwalla and Wilson (1993), too, note that due to the data maturation phenomena, graphical methods of extrapolation based on log–log plot are particularly ill suited. They, instead, suggest that a better approach is dynamic linear models with innovation terms that account for the added uncertainties due to a maturation of the data.

Dynamic linear models. Chen et al. (1996) used different model-year vehicle data for different MIS to forecast warranty claims for a current model-year vehicle at future MIS values using dynamic linear models. In this application, there are certain points to be noted:

- The design and service actions for a previous model-year vehicle may lead to certain specific trends and patterns in the warranty performance. Such features may not be representative of a current model-year vehicle. It is also noted by Singpurwalla and Wilson (1993) that the inclusion of data from previous model years may result in predictions inferior to those given by the current year's data alone.
- There are situations where, for a newly launched vehicle, no previous model-year vehicle data may exist.
- The forecasting results obtained by Chen et al. (1996) are strictly applicable to a given month of interest only. The model does not forecast warranty performance at future months and associated MIS values.

To alleviate the issues faced by the existing methods, in this chapter we develop an artificial neural network black-box model for year-end warranty performance forecasting. We treat warranty performance indicator R/1000 as output and develop a neural network model as an example for R/1000. MIS values and the month at which data gets available are treated as two inputs in each case. The objective is to forecast warranty performance for future months and at associated new MIS values in the presence of the phenomena of maturing data.

In the following sections we study two special types of artificial neural networks: RBFs and MLPs.

9.3 WARRANTY PERFORMANCE FORECASTING USING RBF NEURAL NETWORKS[*]

9.3.1 NETWORK STRUCTURE

Let z_1 denote the number of months since the beginning of the model year of a vehicle and z_2 denote MIS information. The RBF neural network model for R/1000 as output variable is as shown in Figure 9.4.

---

[*] © 2005 Taylor & Francis Group Ltd. This section is reprinted with permission from: Rai, B. K. and Singh, N. (2005). Forecasting warranty performance in presence of maturing data phenomena. *International Journal of System Sciences*, 36(7), 381–394..

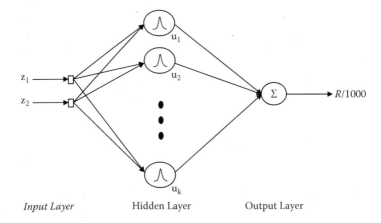

FIGURE 9.4 Radial basis function neural network model for R/1000 forecasting.

As seen in Figure 9.4, a typical RBF neural network has three layers: input layer, hidden layer, and output layer. There are no weights associated with connections between the input layer and the hidden layer. The activation function of each node in the hidden RBF network layer is a Gaussian function and allows nonlinear transformation of the input space to the output space. The activation function u_j determines the output of the hidden layer node and is expressed as

$$u_j = e^{\left[-\frac{\|z-c_j\|^2}{2\sigma_j^2}\right]}, j = 1, 2, \ldots, K \tag{9.1}$$

where, for our example,

$z = \left[z_1, z_2\right]^T$ is the input vector
K is the number of nodes in the hidden layer
c_j represents the cluster center of each hidden layer node $j = 1, 2, \ldots, K$, and its
  dimension is equal to the number of input nodes
σ_j^2 is the radius of cluster for j-th hidden RBF node

The specific region over which a given RBF node has significant activation is decided by c_j and σ_j (Moody and Darken 1989). The exponential decaying localized nonlinearity due to the Gaussian function helps to obtain local approximations to nonlinear input–output mapping. This results in fast learning networks (Haykin 1994).

The transformation from hidden space to output space is linear with output node y for any one-output network given by

$$\hat{y} = \sum_{j=1}^{K} w_j u_j \tag{9.2}$$

where w_j , $j = 1, 2, \ldots, k$ are the set of synaptic weights for connections between the hidden and the output layers.

9.3.2 TRAINING THE RBF NETWORK

Before training of the RBF neural network can begin, one needs to choose the number of hidden nodes k to be used. When the number of hidden nodes is smaller than actually required, the functional approximation may not be very accurate and may result in a network that does not represent the training data (Cook and Chiu 1998). At the same time, too many hidden nodes in the network may result in an overfitting phenomenon. This may produce large testing errors even when very low values for training errors are achieved. Thus, arriving at a suitable number of hidden nodes usually calls for experimentation.

The training scheme for RBF networks involves three main steps (Korres et al. 2002; Xu et al. 2003; Moody and Darken 1989):

Step 1: This step uses the training data set to determine cluster centers c_j for each of the K nodes in the hidden layer. The most common method to achieve this task is the unsupervised K-means clustering approach. This algorithm minimizes the sum of Euclidean distances between each cluster center c_j and the training data points assigned to the particular cluster.

Step 2: In this step, the unit width of each radial unit donated by σ_j is determined. This is achieved using the p-nearest-neighbor heuristic. Two different approaches can be used. In the first approach, for any hidden node j, p-nearest node centers are used for arriving at σ_j using

$$\sigma_j = \sqrt{\left(\frac{1}{p} \sum_{i=1}^{p} \left\| c_j - c_i \right\|^2 \right)} \qquad (9.3)$$

In the second approach, only the p-th nearest neighbor is used for arriving at the value of σ_j. The minimum possible value of p is 1. The maximum value that p can take is K, that is, the number of nodes in the hidden layer. A suitable value for p can be obtained by minimizing training and testing errors through planned experimentation.

Step 3: The previous two steps address nonlinear transformation from input space to hidden space. In this step the connection weight w between the hidden and output layers is obtained through the supervised learning algorithm using the steepest decent method. For a single output network, the mean square error (MSE) given by Equation 9.4 is minimized.

$$MSE = \frac{\sum_{t=1}^{n}(y - \hat{y})^2}{n} \qquad (9.4)$$

where n is the number of training patterns and $\hat{y}$ is the predicted value of y.

Once the RBF neural network is trained using the training data set, its performance is evaluated using the testing data set that is not yet seen by the network. If relatively low error values are obtained for the testing data set, the model can be considered satisfactory.

9.3.3 OPTIMIZATION OF RBF NETWORK STRUCTURE THROUGH EXPERIMENTATION

As discussed in the previous section, arriving at a suitable number of nodes in the hidden layer and a suitable value of p calls for planned experimentation. Table 9.1 gives factors and levels that are chosen for the experimentation (Rai and Singh 2005b).

The performance of the RBF network is known to be very sensitive to the number of nodes in the hidden layer. Thus, in this experiment K is varied in small increments of 2 nodes between 6 and 26. For the second factor P, three levels, that is, ~25% of K, ~50% of K, and ~75% of K, to cover a wide range are chosen. By definition, a smaller value of P will lead to smaller values for σ_j in each hidden node and vice versa. The value of σ_j determines the width of input space to which the j-th hidden node in the RBF neural network responds. The value of σ_j should be large enough so that several nodes in the hidden layer respond with a large output for a given input. On the other hand, the value of σ_j should not be so large that the output of the hidden node does not add any value to the model.

For each warranty performance measure, data points up to month 14 (constituting 105 patterns) are used for training the network. Data points corresponding to month 15 (15 patterns) and month 16 (16 patterns) are used for testing. Thus, out of all the data available up to month 16, about 77% is used for training the network and remaining 23% is used for testing to allow for generalization of the results.

TABLE 9.1

Factors and Levels for Experimentation

Factors	Levels
K: Number of nodes in the hidden layer of RBF network	6, 8, 10, 12, 14, 16, 18, 20, 22, 24, 26
P: Number of nearest neighbors used for determining the width of RBF nodes	Nearest integer values for 25% of K, 50% of K, 75% of K

The output values are normalized based on the training data set. This is done by subtracting each output from the mean of the training data set and then dividing it by corresponding standard deviation. To enable comparison of results, normalized root mean square error (NRMSE) is calculated using the method given by Xu et al. (2003),

$$NRMSE = \sqrt{\frac{\sum_{t=1}^{n}(y-\hat{y})^2}{\sum_{t=1}^{n}y^2}}$$

(9.5)

where n is the number of training/testing patterns and $\hat{y}$ is the predicted value of y.

With factor K having 11 levels and factor P with 3 levels, there are $11 \times 3 = 33$ total possible combinations that can be used for the experimental runs. Since the experimental runs are to be performed on a computer with each run taking only a few seconds, it is decided to run all 33 experimental combinations. The experiments were conducted using BasicRBF program written in the C language (Chinnam 2003). For each experimental combination, two runs each were made using the training data set, and the corresponding NRMSE values were obtained using Equation 9.5. Using the same RBF network structure NRMSE values for the testing data set were also obtained. The full factorial layout and the results obtained for R/1000 are given for both training and testing data sets in Table 9.2.

Using the two NRMSE values corresponding to training and testing data sets for each trial, smaller-the-better type of signal-to-noise ratio (SNR) is calculated using the method given by Taguchi (1992),

$$SNR = -10Log\left[\frac{1}{2}\sum_{i=1}^{2}NRMSE_i^2\right]$$

(9.6)

where $i = 1$ corresponds to the training data set and $i = 2$ corresponds to the testing data set.

By maximizing SNR, suitable levels for K and P can be arrived that minimizes NRMSE values for both the training and testing data sets. The SNR values arrived at using Equation 9.6 are also given in Table 9.2.

Analysis of variance (ANOVA) is carried out using SNR values to understand the impact of K and P on the performance of the RBF network. Table 9.3 shows the ANOVA table.

It is seen from Table 9.3 that the main effects of K and P are statistically significant. The interaction $K \times P$ is significant at the 5% level of significance. To enable choice of the best levels for the two factors K and P, average response graphs are prepared. Average response graphs of main effects and interaction for R/1000 are shown in Figure 9.5.

TABLE 9.2
Full Factorial Experimental Layout and Results

Sl.	K	P	R/1000				SNR-1	SNR-2
			Training		Testing			
			NRMSE-1	NRMSE-2	NRMSE-1	NRMSE-2		
1	6	2	0.333	0.288	0.873	0.448	8.29	19.53
2	6	3	0.278	0.565	0.533	0.569	17.10	11.36
3	6	5	0.235	0.565	0.422	0.569	21.48	11.36
4	8	2	0.252	0.159	0.674	0.531	13.52	18.73
5	8	4	0.072	0.114	0.209	0.446	37.07	22.46
6	8	6	0.069	0.071	0.155	0.158	42.43	42.06
7	10	3	0.149	0.149	0.651	0.420	14.99	23.08
8	10	5	0.081	0.067	0.295	0.297	30.65	30.70
9	10	8	0.058	0.059	0.136	0.160	45.14	42.26
10	12	3	0.141	0.102	0.455	0.480	21.76	21.18
11	12	6	0.061	0.052	0.254	0.280	33.76	32.05
12	12	9	0.053	0.056	0.096	0.187	51.18	39.64
13	14	4	0.102	0.069	0.296	0.461	30.14	22.19
14	14	7	0.052	0.059	0.270	0.319	32.77	29.45
15	14	11	0.192	0.049	0.228	0.222	31.12	36.55
16	16	4	0.069	0.079	0.541	0.389	19.04	25.41
17	16	8	0.043	0.041	0.175	0.197	41.17	39.03
18	16	12	0.094	0.042	0.070	0.058	49.74	59.66
19	18	5	0.052	0.046	0.385	0.394	25.83	25.40
20	18	9	0.047	0.042	0.214	0.118	37.31	48.55
21	18	14	0.078	0.061	0.079	0.141	50.94	44.44
22	20	5	0.040	0.040	0.328	0.367	29.09	26.86
23	20	10	0.045	0.038	0.251	0.216	34.24	37.23
24	20	15	0.218	0.110	0.134	0.236	44.97	33.87
25	22	6	0.033	0.035	0.370	0.340	26.73	28.42
26	22	11	0.068	0.034	0.102	0.269	48.93	33.00
27	22	17	0.405	0.242	0.354	0.094	45.55	33.90
28	24	6	0.034	0.030	0.258	0.299	33.88	30.98
29	24	12	0.205	0.485	0.258	0.168	53.72	29.75
30	24	18	0.209	0.108	0.162	0.184	33.53	37.83
31	26	7	0.030	0.029	0.319	0.367	29.67	26.90
32	26	13	0.180	0.052	0.609	0.370	16.02	26.62
33	26	20	0.137	0.104	0.302	0.176	28.99	38.67

TABLE 9.3

Analysis of Variance for Optimizing the RBF Network Parameters

Source of Variation	Sum of Squares	Degrees of Freedom	Mean Square	F-Ratio	p
K	2770.5	10	277.0	7.4	0.000
P	2711.4	2	1355.7	36.2	0.000
K × P	1422.1	20	71.1	1.9	0.050
Error	1237.0	33	37.5		
Total	8140.9	65			

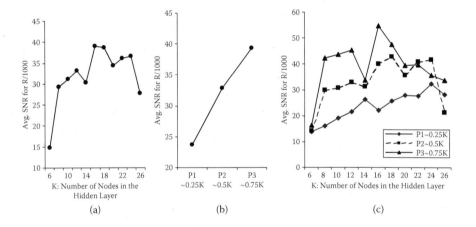

(a)   (b)   (c)

FIGURE 9.5   Average response graphs for R/1000.

Figure 9.5a shows that when the number of nodes in the hidden layer is increased, SNR value increases and achieves maximum value at $K = 16$ and then decreases. Average response graph for main effect P in Figure 9.5b shows linearly increasing trend when the number of nearest neighbors used is changed from ~25% of K to ~75% of K. This result indicates that better results from the RBF network are attained when higher number of p-nearest neighbors is used. Figure 9.5c shows the interaction average response graph for $K \times P$. It can be seen that across different levels of factor K, third level of factor P gives overall better performance. Maximizing SNR is equivalent to minimizing NRMSE values together for both the training and testing data sets. The best levels chosen for K and P are 16 and 12, respectively.

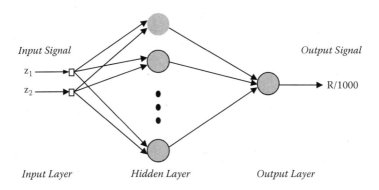

Input Signal

z_1

z_2

Output Signal

R/1000

Input Layer *Hidden Layer* *Output Layer*

FIGURE 9.6 MLP network for R/1000 forecasting.

9.4 FORECASTING WARRANTY PERFORMANCE USING MLP NEURAL NETWORKS*

9.4.1 NETWORK STRUCTURE

The MLP network architecture for R/1000 as output variable is shown in Figure 9.6. It shows an MLP network with two input signals in the input layer, a single layer of hidden neurons, and an output layer with a single response. It can also be seen that a neuron in any layer of the network is connected to all the nodes or neurons in the previous layer. There are two types of signals that flow through the MLP network, such as function signals and error signals. Figure 9.7 depicts the two types of signals.

A function signal begins at the input end of the network, moves from left to right from one layer to another, and ends at the output end of the network. On the other hand, an error signal originates at the output end of the network and moves backward from right to left through the network. The error signal at the output of neuron j at iteration n is defined as

$$e_j(n) = y_j(n) - \hat{y}_j(n) \qquad (9.7)$$

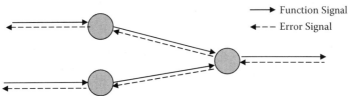

⟶ Function Signal
◀--- Error Signal

FIGURE 9.7 Two types of signals in the MLP network.

* © 2005 Springer. With kind permission from Springer Science & Business Media: Rai, B. K. and Singh, N. (2005). Forecasting automobile warranty performance in presence of maturing data phenomena using multilayer perceptron neural network. *Journal of System Science and Systems Engineering,* 14(2), 159–176.

where $y_j(n)$ represents the desired response of neuron j for iteration n, and $\hat{y}_j(n)$ represents the output of neuron j at iteration n.

The MLP network is trained using a training data set. The number of neurons in the hidden layer is often arrived through experimentation. During the training phase, the values of synaptic weights in the network are optimized by minimizing the mean square error given by Equation 9.4.

Once an MLP network is trained using the training data, its performance is evaluated using a testing data set that is not yet seen by the network. If relatively low error values are obtained for the testing data set, the network model is considered satisfactory.

9.4.2 OPTIMIZATION OF MLP NETWORK USING RESPONSE SURFACE METHODOLOGY

There are several parameters of an MLP network that influence its performance, such as number of hidden layers, number of neurons in each layer, learning rate, momentum, training mode, etc. Experiments are conducted using an MLP network with a single layer of hidden neurons. Two control factors chosen for experimentation are the number of neurons in the hidden layer (N) and learning rate (L). Central composite design (CCD) is chosen as the experimental design (Myers and Montgomery 2002). The control factors and their levels chosen for the experimentation are given in Table 9.4 (Rai and Singh 2005c). Table 9.5 shows the design matrix with coded (x_1 and x_2) and physical (N and L) levels. The graphical depiction of the design is shown in Figure 9.8.

Figure 9.8 shows that the design consists of eight equally spaced points on a circle of radius 1.414 and five runs at the center. The four design points (0, 1.414), (0, −1.414), (1.414, 0), and (−1.414, 0) are called *axial points* as they lie on the x_1 and x_2 axes. The points allow for estimation of pure quadratic terms, such as x_1^2 and x_2^2. These points, however, are not designed to obtain interaction term $x_1 x_2$.

For training the network, this study involves R/1000 values at each MIS, available through the first 14 months of a given model-year vehicle. The z_1, z_2, and corresponding R/1000 values for the first 14 months thus constitute

$$105 \left(= \frac{14 \times 15}{2} \times 100 \right)$$

training patterns. The R/1000 values for the 15th and 16th month with 15 and 16 patterns, respectively, together constitute 31 patterns for testing the MLP network. Thus, about

$$77\% \left(= \frac{105}{(105+31)} \times 100 \right)$$

of the R/1000 values are used for training the network, and the remaining

TABLE 9.4

Factors and Levels for the Central Composite Design

Factors	Levels				
	−1.414	−1	0	+1	+1.414
N: Number of neurons in the hidden layer	11	15	25	35	39
L: Learning rate	0.0293	0.05	0.1	0.15	0.1707

TABLE 9.5

Design Layout for the Experiment

X_1	X_2	N	L
−1	−1	15	0.0001
1	−1	35	0.0001
−1	1	15	0.0005
1	1	35	0.0005
−1.41421	0	11	0.0003
1.41421	0	39	0.0003
0	−1.41421	25	0.0000172
0	1.41421	25	0.0005828
0	0	25	0.0003
0	0	25	0.0003
0	0	25	0.0003
0	0	25	0.0003
0	0	25	0.0003

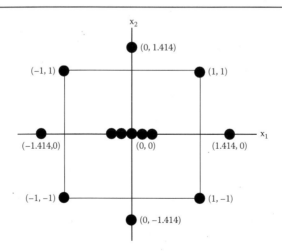

FIGURE 9.8   Graphical representation of the experimental design.

$$23\% \left(= \frac{31}{(105+31)} \times 100 \right)$$

are used for testing to allow for generalization.

Before training the network, the R/1000 values are normalized by subtracting the mean of the data set from each value and then dividing it by the corresponding standard deviation. To enable comparison of results between the two warranty performance indicators, NRMSE is calculated using Equation 9.5.

An MLP network trained twice using the same parameters can result in different NRMSE values due to use of random seeds in assigning initial values to the synaptic weights in the network. For this reason, for a given set of network parameters, two NRMSE values each are obtained, based on the training and testing data sets. As NRMSE is a smaller-the-better type of characteristic, the corresponding SNR is calculated using Equation 9.6.

By maximizing the SNR, suitable levels for N and L can be arrived at that minimizes NRMSE values for both training as well as testing data sets. The factors kept constant during the experimentation are

Number of layers with hidden neurons: 1
Activation function: Hyperbolic tangent (a = 1.7159; b = 2/3)
Momentum: 0
Acceptable training error per node: 0.00000001
Maximum number of epochs: 20,000

The NRMSE and SNR values obtained from the experiment are given in Table 9.6.

The results from the experiment were analyzed using Minitab software. The quadratic term for factor L was found to be nonsignificant and was thus removed from the model. The estimated coefficients for the remaining terms are given in Table 9.7, and the ANOVA table for SNR is given in Table 9.8.

Table 9.7 shows that the lack-of-fit term is not significant. Figure 9.9 shows the residual plots, which show a satisfactory pattern. Figure 9.10 shows the contour plot of SNR.

Figure 9.10 shows that the higher values of SNR would be obtained in the southeast direction. It also suggests that additional experimentation to improve upon the results would lie outside the current experimental region. Based on the second combination in Table 9.6, that yields an SNR of 61.54 dB; $N = 35$ and $L = 0.0001$ were chosen as best levels.

TABLE 9.6
Physical Layout with NRMSE and SNR Values

N: Number of Neurons	L: Learning Rate	NRMSE-1		NRMSE-2		SNR
		(Trg)	(Test)	(Trg)	(Test)	
15	0.0001	0.041	0.116	0.042	0.082	51.31
35	0.0001	0.034	0.064	0.035	0.045	61.54
15	0.0005	0.043	0.114	0.045	0.053	53.15
35	0.0005	0.040	0.109	0.046	0.087	51.51
11	0.0003	0.041	0.080	0.038	0.114	51.83
39	0.0003	0.037	0.088	0.037	0.080	54.74
25	0.0000172	0.034	0.086	0.037	0.091	53.96
25	0.0005828	0.037	0.111	0.065	0.132	47.31
25	0.0003	0.037	0.114	0.043	0.125	48.31
25	0.0003	0.046	0.090	0.046	0.113	50.68
25	0.0003	0.039	0.140	0.037	0.080	49.31
25	0.0003	0.050	0.121	0.036	0.077	51.06
25	0.0003	0.037	0.078	0.037	0.106	52.99

TABLE 9.7
Regression Coefficients in Coded Units

Term	Coef	SE Coef	T	P
Constant	50.961	0.7071	72.068	0.000
X_1	1.575	0.6922	2.275	0.052
X_2	−2.197	0.6922	−3.174	0.013
$X_1 * X_1$	1.901	0.7360	2.583	0.032
$X_1 * X_2$	−2.975	0.9790	−3.039	0.016

$S = 1.958 \ R\text{-}Sq = 79.6\% \ R\text{-}Sq(adj) = 69.4\%$

9.5 RESULTS AND DISCUSSIONS*

For the RBF network, the best values of $K = 16$ and $P = 12$ were arrived at by minimizing the training and the testing errors together. From Table 9.2 we observe that the least values of NRMSE for the training data set is attained in the 31st row when $K = 26$ and $P = 7$. The minimum NRMSE values based on the training data set for R/1000 is 0.029, an improvement of approximately 31% when compared to the

---

* © 2005 Springer. With kind permission from Springer Science & Business Media: Rai, B. K. and Singh, N. (2005). Forecasting automobile warranty performance in presence of maturing data phenomena using multilayer perceptron neural network. *Journal of System Science and Systems Engineering,* 14(2), 159–176.

TABLE 9.8

Analysis of Variance for SNR

Source	DF	Seq SS	Adj SS	Adj MS	F	P
Regression	4	119.44	119.44	29.860	7.79	0.007
Linear	2	58.46	58.46	29.230	7.62	0.014
Square	1	25.58	25.58	25.577	6.67	0.032
Interaction	1	35.40	35.40	35.403	9.23	0.016
Residual error	8	30.67	30.67	3.834		
Lack-of-fit	4	17.74	17.74	4.435	1.37	0.383
Pure error	4	12.93	12.93	3.232		
Total	12	150.11				

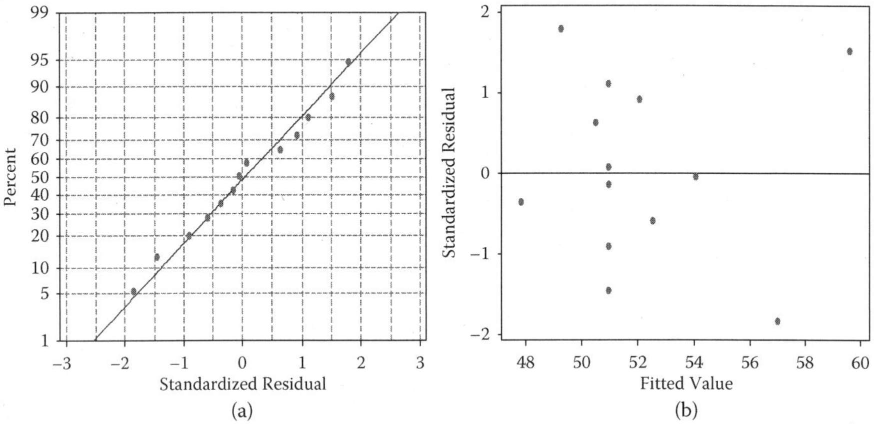

FIGURE 9.9   Residual plots: (a) normal probability plot of residuals, (b) residual versus fitted values.

NRMSE values for the best runs identified. However, corresponding NRMSE values for the testing data set show an opposite trend. The NRMSE values based on the testing data set corresponding to the minimum training errors for R/1000 is 0.367, an increase in the NRMSE value by 6.3 times as compared to the best runs identified.

It is also observed from Table 9.6 that low NRMSE values for the training data set not always leads to low NRMSE values for the testing data set. Based on training and testing errors being high or low, there are four scenarios that can occur as depicted in Figure 9.11.

In Figure 9.11a, both training and testing errors are low. This implies that a good generalization is achieved from the trained MLP network. Figure 9.11b shows low training errors but high testing errors. This can occur due to overfitting of the training data set. It is known that, by increasing the number of neurons in the hidden layer, often one can achieve lower errors values for the training data set. However,

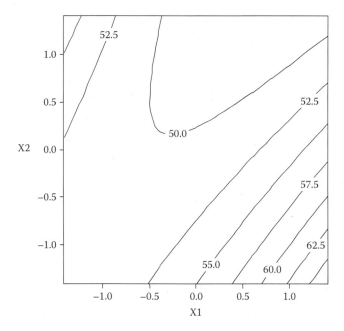

FIGURE 9.10   Contour plot of SNR.

such low error values, do not necessarily translate to low error values for the testing data set as seen in this example. Figure 9.11c shows a situation where although training errors are high, the testing errors are low. Such a situation may occur only by a remote chance. Such network should not be used for forecasting, as there is no guarantee of repeatability. Figure 9.11d shows a situation where both training and testing errors are high. With high training errors, it is natural to expect high testing errors.

A comparison of NRMSE values obtained from three methods, such as log–log plot, MLP networks, and an RBF network to forecast R/1000 values are given in Figure 9.12.

Figure 9.12 shows NRMSE values based on R/1000 values forecasted 9 months into the future beyond the training data set. The NRMSE values are observed to be based on log–log plots, and the MLP network increase as the forecasting horizon increases. NRMSE values based on the MLP network show a much better overall performance as compared to the log–log plot method and RBF networks. NRMSE obtained from the RBF network shows the lowest values for months 15 and 16 as compared to the other two methods. However, thereafter, it increases at a faster rate than that compared to NRMSE values from the log–log plots and MLP network. The results also suggest that, for short-term forecasting (say 1 to 2 months) of R/1000 values, RBF networks have a slight edge over MLP networks. However, for longer-term forecasting, say, 3 to 9 months into the future, the MLP network provides a much better forecasting performance. Figure 9.13 takes a closer look at the performance of the three methods.

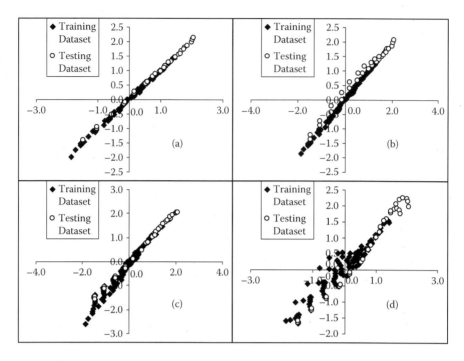

FIGURE 9.11   Scatter plot of normalized actual versus predicted values for R/1000. (a) Both training and testing errors are low [Trg. R-sq = 0.9983, Testing R-sq = 0.9985], (b) low training error, high testing error [Trg. R-sq = 0.998, Testing R-sq = 0.963], (c) high training error, low testing error [Trg. R-sq = 0.963, Testing R-sq = 0.997], (d) high training error, high testing error [Trg. R-sq = 0.913, Testing R-sq = 0.974].

Figure 9.13 shows observed versus predicted R/1000 values after 17 and 18 months from the start of the model year for the three methods. Months 17 and 18 correspond to the 3rd and 4th month, respectively, in future beyond the training data set. The predicted R/1000 based on the log–log plots clearly shows maximum departure from the actual R/1000 values. The departure from the observed R/1000 values initiate at about 9 MIS point and reaches a maximum at 17 and 18 MIS for months 17 and 18, respectively. Since log–log plots use the latest available data, which in this case is for month 13, and do not take into account warranty growth phenomena, the underestimation is an obvious outcome.

For prediction using the RBF network, the departure from the observed R/1000 values start at about 14 MIS and reaches a maximum at 17 and 18 MIS for months 17 and 18, respectively. It is to be noted that R/1000 values at different MIS within each month from the start of the model-year vehicle increase with increasing MIS value. This is due to cumulative nature of the R/1000 value. This implies that R/1000 values beyond the training data set at new MIS would continue to increase as compared to R/1000 at preceding MIS points.

Xu et al. (2003) observe that the ability of the RBF network to recognize whether an input is near the training set or if it is an untrained region of the input space gives

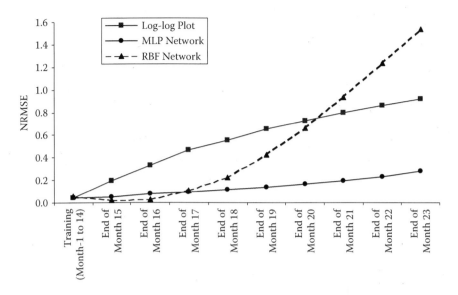

FIGURE 9.12 Comparison of NRMSE values obtained from use of log–log plot method, MLP networks, and RBF networks to forecast R/1000.

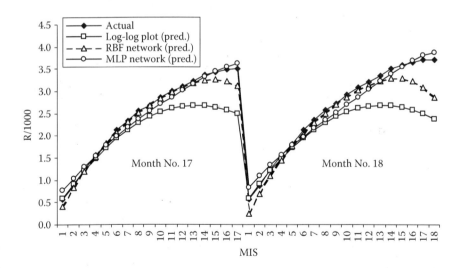

FIGURE 9.13 Observed versus predicted R/1000 for the log–log plot method, RBF networks, and MLP networks at months 17 and 18.

the RBF a significant advantage over the MLP network structure. However, the same ability turns into a disadvantage when forecasting values beyond the range seen by the network, which is required when forecasting R/1000 values. Vojinovic et al. (2001) and Wedding and Cios (1996) note that RBF networks are very poor at extrapolating an answer when the input data are considerably different from what they were trained on. Wedding and Cios (1996) further elaborate that, in the case of Gaussian functions, as the input vectors get farther away from the center of a cluster, the resulting output gets exponentially smaller and quickly approaches zero. This implies that RBFs using Gaussian transforms are good only when the input data are near or within the RBF centers. Thus, MLP networks that overcome the inherent weaknesses of the log–log plot method, and RBF networks, are expected to perform better when predicting R/1000, which is expected to be higher at future MIS and at future months.

R/1000 values forecasted with the optimal MLP network show a good fit for different MIS values for both the 17th and 18th month. Figure 9.14 shows the observed versus predicted R/1000 values based on the optimized MLP network. Note that the data in Figure 9.14 are similar to the data in Figure 9.2, except that, rather than overlaying the curves, they appear end to end.

It can be seen from Figure 9.14 that the MLP network based on R/1000 values from the first 14 months of a given model-year vehicle forecasts R/1000 values at various MIS beyond the month 14 fairly accurately. However, it is to be noted that, as the forecasting horizon increases to about 8 or more months in the future, the forecasting error starts to become larger.

It is shown that a sound forecasting model to predict year-end warranty performance can be successfully developed in the presence of the phenomena of maturing data. The forecasting model developed using the MLP network can help the warranty cost reduction team of engineers to forecast R/1000 values into the future 6 to 7 months fairly accurately and fine-tune their strategies to achieve year-end targets. When new data become available, the existing forecasting model can be updated by restricting the search for the best levels of N and L in the neighborhood of those obtained in this article.

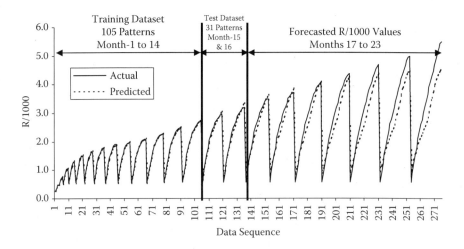

FIGURE 9.14 Observed versus predicted R/1000 for the training data set, testing data set, and forecasted values from the MLP neural network.

9.6 SUMMARY

Key points from this chapter are summarized here.

- Neural networks provide an effective tool for forecasting warranty performance in the presence of warranty growth or maturing data issue associated with warranty data. Design of experiments and response surface methodology provide useful directions for optimizing the neural network parameters. Experimental design methods help to identify the key variables for a given neural network and also suggest certain ranges for the variables that can yield better performance. Thereafter, training and validating the network model based on satisfactory results can optimize neural network parameters.
- For short-term forecasting of warranty performance that involves prediction up to 3 months into the future, both RBF and MLP neural networks have comparable performances. Both show significantly better forecasting performance as compared to the log–log plot method.
- Forecasting warranty performance that involves 4 months and beyond into the future, MLP neural networks significantly outperform RBF networks and the log–log plot method. Due to the cumulative nature of the warranty performance indicators, RBF networks are ill suited for forecasting beyond 3 months into the future as data points beyond those originally seen by the network are encountered. Since MLP networks perform satisfactorily for both short-term and longer-term forecasting of warranty performance, their use is recommended.

BIBLIOGRAPHIC NOTES

Neural networks are extensively used for a variety of forecasting problems. Wedding and Cios (1996) used RBF neural networks for time-series forecasting by combining the method with the univariant Box–Jenkins model. Ho et al. (2002) carried out a comparative study of neural networks and the Box–Jenkins autoregressive integrated moving average (ARIMA) models in predicting failures of repairable compressor systems. Korres et al. (2002) proposed an RBF neural network model to approximate fuel lubricity level using other fuel properties as inputs. Yao et al. (2002) developed models for predicting the critical temperature of 856 organic compounds using multiple linear regression and RBF neural networks based on descriptors calculated from molecular structure. Xu et al. (2003) applied neural networks to forecast engine system reliability. Zemouri et al. (2003) successfully applied recurrent RBF neural networks on three different time-series prediction problems. Rank (2003) used a Bayesian-trained RBF network for regenerating time series generated by the Lorenz system and speech signals. Anijdan and Bahrami (2005) used several different network architectures to develop a model for prediction of topologically closed-packed phase formation in superalloys and reported MLP network to possess most attractive properties in terms of performance. Riad et al. (2004) used multilayer perceptron neural network to model the rainfall-runoff relationship in a catchment located in a semiarid climate in Morocco and predict river runoff. Gaudart et al. (2004) compared the prediction and estimation performance of multilayer perceptron and linear regression for epidemiological data. Hutchinson (1994) used RBF networks for stock price prediction.

References

Andrews, J. D. and Moss, T. R. (2002). *Reliability and Risk Assessment*. The American Society of Mechanical Engineers, New York.

Anijda, S. H. M. and Bahrami, A. (2005). A new method in prediction of TCP phases formation in superalloys. *Materials Science and Engineering A*, 396, 138–142.

Baik J., Murthy, D. N. P., and Jack, N. (2004). Two-dimensional failure modeling with minimal repair. *Naval Research Logistics*, 51, 1–18.

Blischke, W. R. and Murthy, D. P. N. (1994). *Warranty Cost Analysis*. New York: Marcel Dekker.

Blischke, W. R. and Murthy, D. P. N. (2000). *Reliability Modeling, Prediction, and Optimization*. New York: John Wiley & Sons.

Chen, J., Lynn, N. J., and Singpurwalla, N. D. (1996). Forecasting warranty claims. In *Product Warranty Handbook*, R. Blischke and D. P. N. Murthy, Eds. New York: Marcel Dekker.

Chen, T. and Popova, E. (2002). Maintenance policies with two-dimensional warranty. *Reliability Engineering and System Safety*, 77, 61–69.

Chinnam, R. B. (2003). BasicRBF program written in C language. Personal Communication, Industrial and Manufacturing Engineering Department, Wayne State University, Detroit, MI.

Cohen, C. A. (1949). On estimating the mean and standard deviation of truncated normal distributions. *Journal of American Statistical Association*, 44, 518–525.

Cohen, C. A. (1951). On estimating the mean and variance of singly truncated normal frequency distributions from the first three sample moments. *Annals of the Institute of Statistical Mathematics*, 3, 37–44.

Cohen, C. A. (1957). On the solution of estimating equations for truncated and censored samples from normal populations. *Journal of American Statistical Association*, 44, 225–236.

Cohen, C. A. (1959). Simplified estimators for the normal distribution when samples are singly censored or truncated. *Technometrics*, 1, 217–237.

Cohen, C. A. (1991). *Truncated and Censored Samples: Theory and Applications*. New York: Marcel Dekker.

Cohen, C. A. and Whitten, B. J. (1988). *Parameter Estimation in Reliability and Life Span Models*. New York: Marcel Dekkar.

Cook, D. F. and Chiu, C. C. (1998). Using radial basis function neural networks to recognize shifts in correlated manufacturing process parameters. *IIE Transactions*, 30, 227–234.

Crowder, M. J., Kimber, A. C., Smith, R. L., and Sweeting, T. J. (1991). *Statistical Analysis of Reliability Data*. London: Chapman & Hall.

Davis, T. P. (2003). *Global Vehicle Reliability—Prediction and Optimization Techniques*. J. E. Strutt and P. L. Hall, Eds. London: Professional Engineering Publishing.

Deming, W. E. (2000). *The New Economics for Industry, Government, Education*. Massachusetts: The MIT Press.

Elkins, D. A. and Wortman, M. A. (2004). An examination of correlation effects among warranty claims. *IEEE Transactions on Reliability*, 53(2), 200–204.

Fisher, R. A. (1931). Properties and applications of Hh functions. *Introduction to British AAS Mathematical Tables*, 1, xxvi–xxxv.

Galton, F. (1897). An examination into the registered speeds of American trotting horses with remarks on their value as hereditary data. *Proceedings of the Royal Society of London*, 62, 310–314.

Gaudart, J., Giusiano, B., and Huiart, L. (2004). Comparison of the performance of multilayer perceptron and linear regression for epidemiological data. *Computational Statistics and Data Analysis*, 44, 547–570.

Goel, P. S. and Singh, N. (1999). A new paradigm for durable product design with up-front consideration of lifetime performance. *Robotics and Computer-Integrated Manufacturing*, 15, 65–75.

Gupta, A. K. (1952). Estimation of the mean and standard deviation of a normal population from a censored sample. *Biometrika*, 39, 260–273.

Gupta, S. C. and Kapoor, V. K. (1989). *Fundamentals of Mathematical Statistics*. New Delhi: Sultan Chand and Sons.

Harry, M. and Schroeder, R. (2000). *Six Sigma: The Breakthrough Management Strategy Revolutionizing the World's Top Corporations*. New York: Currency.

Haykin, S. (1994). *Neural Networks: A Comprehensive Foundation*. Macmillan College Publishing Company.

Hill, V. L., Beall, C. W., and Blischke, W. R. (1991). A simulation model for warranty analysis. *International Journal of Production Economics*, 22, 131–140.

Ho, S. L., Xie, M., and Goh, T. N. (2002). A comparative study of neural network and Box-Jenkins ARIMA modeling in time series prediction. *Computers and Industrial Engineering*, 42, 371–375.

Hoyland, A. and Rausand, M. (1994). *System Reliability Theory—Models and Statistical Methods*. New York: John Wiley & Sons.

Hu, X. J. and Lawless, J. F. (1996). Estimation of rate and mean functions from truncated recurrent event data. *Journal of American Statistical Association*, 91, 300–310.

Hu, X. J., Lawless, J. F., and Suzuki, K. (1998). Nonparametric estimation of a lifetime distribution when censoring times are missing. *Technometrics*, 40, 3–13.

Hutchinson, J. M. (1994). A Radial Basis Function Approach to Financial Time Series Analysis. Ph.D. thesis, MIT.

Inman, R. R. and Gonsalvez, D. J. A. (1998). A cost-benefit model for production vehicle testing. *IIE Transactions*, 30, 1153–1160.

Iskandar, B. P. and Blischke, W. R. (2003). Reliability and warranty analysis of a motorcycle based on claims data, in *Case Studies in Reliability and Maintenance*, W. R. Blischke and D. N. P. Murthy, Eds. New Jersey: Wiley-Interscience, pp. 623–656.

Kakouros, S., Cargille, B., and Esterman, M. (2003). Reinventing warranty at HP: An engineering approach to warranty. *Quality and Reliability Engineering International*, 19, 21–30.

Kalbfleisch, J. D. and Lawless, J. F. (1988). Estimation of reliability in field-performance studies. *Technometrics*, 30, 365–388.

Kalbfleisch, J. D. and Lawless, J. F. (1992). Some useful statistical methods for truncated data. *Journal of Quality Technology*, 24, 145–152.

Kalbfleisch, J. D., Lawless, J. F., and Robinson, J. A. (1991). Methods for the analysis and prediction of warranty claims. *Technometrics*, 33, 273–285.

Kaplan, E. L. and Meier, P. (1958). Nonparametric estimation from incomplete observations. *Journal of the American Statistical Association*, 53, 457–481.

Karim, M. R., Yamamoto, W., and Suzuki, K. (2001). Analysis of marginal count failure data. *Lifetime Data Analysis*, 7(2), 173–186.

Kececioglu, D. (1993). *Reliability and Life Testing Handbook*, Vol. 1. New Jersey: PTR Prentice Hall.

Korres, D. M., Anastopoulos, G., Lois, E., Alexandridis, A., Sarimveis, H., and Bafas, G. (2002). A neural network approach to the prediction of diesel fuel lubricity. *Fuel*, 81, 1243–1250.

Krivstov, V. V., Tanako, D. E., and Davis, T. P. (2002). Regression approach to tire reliability analysis. *Reliability Engineering and System Safety*, 78, 267–273.

Lawless, J. F. (1982). *Statistical Models and Methods for Lifetime Data*. New York: John Wiley & Sons.

Lawless, J. F. (1987). Regression methods for Poisson process data. *Journal of the American Statistical Association*, 82, 808–815.

Lawless, J. F. (1995). The analysis of recurrent events for multiple subjects. *Applied Statistics*, 44, 487–498.

Lawless, J. F. and Nadeau, C. (1995). Some simple robust methods for the analysis of recurrent events. *Technometrics*, 37, 158–168.

Lawless, J. F., Hu, J., and Cao, J. (1995). Methods for the estimation of failure distributions and rates from automobile warranty data. *Lifetime Data Analysis*, 1, 227–240.

Lu, M. W. (1998). Automotive reliability prediction based on early field failure warranty data. *Quality and Reliability Engineering International*, 14, 103–108.

Majeske, K. D. (2003). A mixture model for automobile warranty data. *Reliability Engineering and System Safety*, 81, 71–77.

Majeske, K. D., Lynch-Caris, T., and Herrin, G. (1997). Evaluating product and process design changes with warranty data. *International Journal of Production Economics*, 50, 79–89.

Meeker, W. Q. and Escobar, L. A. (1998). *Statistical Methods for Reliability Data*. New York: John Wiley & Sons.

Meeker, W. Q. and Escobar, L. A. (2004). Reliability: The other dimension of quality. *Quality Technology and Quantitative Management*, 1(1), 1–25.

MIL-STD 882. (1984). System Safety Program Requirement. U.S. Department of Defense, Washington, DC.

Montgomery, D. C. (1991). *Design and Analysis of Experiments*. New York: John Wiley & Sons.

Moody, J. and Darken, C. J. (1989). Fast learning in networks of locally tuned processing units. *Neural Computation*, 1, 281–294.

Murthy, D. N. P. and Djamaludin, I. (2002). New product warranty: A literature review. *International Journal of Production Economics*, 79, 231–260.

Murthy, D. N. P, Xie, M., and Jiang, R. (2004). *Weibull Models*. Hoboken, New Jersey: Wiley-Interscience: A John Wiley & Sons.

Myers, R. H. and Montgomery, D. C. (2002). *Response Surface Methodology: Process and Product Optimization using Designed Experiments*. New York: John Wiley & Sons.

Nelson, W. (1982). *Applied Life Data Analysis*. New York: John Wiley & Sons.

Nelson, W. (1988). Graphical analysis of system repair data. *Journal of Quality Technology*, 20, 24–35.

Nelson, W. (1990). Hazard plotting of left truncated life data. *Journal of Quality Technology*, 22, 230–238.

Nelson, W. (1995). Confidence limits for recurrence data applied to cost or number of product repairs. *Technometrics*, 37, 147–157.

Nelson, W. (2000). Theory and applications of hazard plotting for censored failure data. *Technometrics*, 42, 12–25.

Nelson, W. (2003). *Recurrent Events Data Analysis for Product Repairs, Disease Recurrences, and Other Applications*. SIAM: Philadelphia, Pennsylvania.

Oh, Y. S. and Bai, D. S. (2001). Field data analysis with additional after-warranty failure data. *Reliability Engineering and System Safety*, 72, 1–8.

Pal, S. and Murthy, G. S. R. (2003). An application of Gumbel's bivariate exponential distribution in estimation of warranty cost of motor cycle. *The International Journal of Quality and Reliability Management*, 20(4), 488–502.

Pearson, K. (1902). On the systematic fitting of frequency curves. *Biometriks*, 2, 2–7.

Pearson, K. and Lee, A. (1908). On the generalized probable error in multiple normal correlation. *Biometriks*, 6, 59–69.

Petkova, V. T., Sander, P. C., and Brombacher, A. C. (1999). The role of the service center in improvement processes. *Quality and Reliability Engineering International*, 15, 431–437.

Petkova, V. T., Sander, P. C., and Brombacher, A. C. (2000). The use of quality metrics in service centers. *International Journal of Production Economics*, 67, 27–36.

Phadke, M. S. (1989). *Quality Engineering Using Robust Design*. New Jersey: Prentice Hall.

Rai, B. K. (2004). Modeling, Analysis and Prediction from Automobile Warranty Datasets. Ph.D. thesis, Wayne State University.

Rai, B. K. and Singh, N. (2003a). Hazard rate estimation from incomplete and unclean warranty data. *Reliability Engineering and System Safety*, 81, 79–92.

Rai, B. K. and Singh, N. (2003b). Developing hazard plots from truncated warranty datasets. Conference proceedings of *Canadian Reliability and Maintainability Symposium 2003*, P52, 1–22.

Rai, B. K. and Singh, N. (2004a). Modeling and analysis of automobile warranty data in presence of bias due to customer-rush near warranty expiration limit. *Reliability Engineering and System Safety*, 86, 83–94.

Rai, B. K. and Singh, N. (2004b). Nonparametric estimation of hazard rate from incomplete, unclean, and biased automobile warranty data. Conference proceedings of *Industrial Engineering Research Conference*, Houston.

Rai, B. K. and Singh, N. (2005a). A modeling framework for assessing the impact of new time/mileage warranty limits on the number and cost of automotive warranty claims. *Reliability Engineering and System Safety*, 88(2), 157–169.

Rai, B. K. and Singh, N. (2005b). Forecasting warranty performance in presence of "maturing data" phenomena. *International Journal of System Sciences*, 36(7), 381–394.

Rai, B. K. and Singh, N. (2005c). Forecasting automobile warranty performance in presence of "maturing data" phenomena using multilayer perceptron neural network. *Journal of System Science and Systems Engineering*, 14(2), 159–176.

Rai, B. K. and Singh N. (2006). Customer-rush near warranty expiration limit and nonparametric hazard rate estimation from known mileage accumulation rates. *IEEE Transactions on Reliability*, 55(3), 480–489.

Raj, D. (1952). On estimating the parameters of normal populations from singly truncated samples. *Ganita*, 3, 41–57.

Rank, E. (2003). Application of Bayesian trained RBF networks to nonlinear time-series modeling. *Signal Processing*, 83, 1393–1410.

Riad, S., Mania, J., Bouchaou, L., and Najjar, Y. (2004). Rainfall-runoff model using an artificial neural network approach. *Mathematical and Computer Modeling*, 40, 839–846.

Robinson, J. A., and McDonald, G. C. (1990). Issues related to field reliability and warranty data. In *Data Quality Control—Theory and Pragmatics*, G. E. Liepins and V. R. R. Uppuluri, Eds. New York: Marcel Dekker, pp. 69–90.

Sander, P. C., Toscano, L. M., Luitjens, S., Petkova, V. T., Huijben, A., and Brombacher, A. C. (2003). Warranty data analysis for assessing product reliability. In *Case Studies in Reliability and Maintenance*, W. R. Blischke and D. N. P. Murthy, Eds. New Jersey: Wiley-Interscience, pp. 591–621.

Shi, Z., Tamura, Y., and Ozaki, T. (1999). Nonlinear time series modeling with the radial basis function-based state-dependent autoregressive model. *International Journal of Systems Science*, 30(7), 717–727.

Singpurwalla, N. D. and Wilson, S. (1993). The warranty problem: its statistical and game theoretic aspects. *SIAM Review*, 35, 17–42.

Stephens, D. and Crowder, M. (2004). Bayesian analysis of discrete time warranty data. *Applied Statistics*, 53, 195–217.

Steve, S. (1976). A graphical estimate of the population mean from censored normal data. *Applied Statistics*, 25, 8–11.

Stevens, W. L. (1937). The truncated normal distribution. (Appendix to paper by CI Bliss, The calculation of the time mortality curve.) *Annals of Applied Biology*, 24, 815–852.

Suzuki, K. (1985a). Nonparametric estimation of lifetime distributions from a record of failures and follow-ups. *Journal of American Statistical Association*, 80, 68–72.

Suzuki, K. (1985b). Estimation of lifetime parameters from incomplete field data. *Technometrics*, 27, 263–271.

Suzuki, K. (1995). Role of field performance data and its analysis. *Recent Advances in Life-Testing and Reliability*, 141–151.

Suzuki, K., Md. Karim, R., and Wang, L. (2001). Statistical analysis of reliability warranty data. In *Handbook of Statistics*, N. Balakrishnan and C. R. Rao, Eds. Amsterdam: Elsevier Science, chap. 21.

Swamy, P. S. (1962). On the joint efficiency of the estimates of the parameters of normal populations based on singly and doubly truncated samples. *Journal of American Statistical Association*, 57, 46–53.

Taguchi, G. (1992). *Taguchi Methods: Research and Development*. ASI Press, Dearborn.

Turnbull, B. W. (1974). Nonparametric estimation of a survivorship function with doubly censored data. *Journal of American Statistical Association*, 69, 169–173.

Vojinovic, Z., Kecman, V., and Seidel, R. (2001). A data mining approach to financial time series modeling and forecasting. *International Journal of Intelligent Systems in Accounting, Finance and Management*, 10, 225–239.

Wedding, D. K., and Cios, K. J. (1996). Time series forecasting by combining RBF networks, certainty factors, and the Box-Jenkins model. *Neurocomputing*, 10, 149–168.

Wu, H., and Meeker, W. Q. (2002). Early detection of reliability problems using information from warranty databases. *Technometrics,* 44, 20–33.

Wu, Y. and Wu, A. (2000). *Taguchi Methods for Robust Design*. New York: ASME.

Xu, K., Xie, M., Tang, L. C., and Ho, S. L. (2003). Application of neural networks in forecasting engine systems reliability. *Applied Soft Computing*, 2, 255–268.

Yadav, O. P., Singh, N., Chinnam, R. B., and Goel, P. S. (2003). A fuzzy logic based approach to reliability improvement estimation during product development. *Reliability Engineering and System Safety*, 80, 63–74.

Yao, X., Wang, Y., Zhang, X., Zhang, R., Liu, M., Hu, Z., and Fan, B. (2002). Radial basis function neural network-based QSPR for the prediction of critical temperature. *Chemometrics and Intelligent Laboratory Systems*, 62, 217–225.

Zakkula, G. (1966). Characterization of normal and generalized truncated normal distributions using order statistics. *Annals of Mathematical Statistics*, 37, 1011–1015.

Zemouri, R., Racoceanu, D., and Zerhouni, N. (2003). Recurrent radial basis function network for time-series prediction. *Engineering Applications of Artificial Intelligence*, 16, 453–463.

Index